Healing Through
SOUL
INTELLIGENCE®

Kristine Genovese

Published by

The Los Angeles Tribune Publishing House

611 Wilshire Blvd, Suite 900, Los Angeles, CA 90017

Printed in the United States of America

The information in this book is intended for informational purposes only. The author and publisher make no representations or warranties with respect to the accuracy or completeness of the contents of this book and specifically disclaim all warranties, including without limitation warranties of fitness for a particular purpose. The advice and strategies contained herein may not be suitable for every situation. Readers should seek the services of a qualified professional where appropriate. Neither the author nor the publisher shall be liable for damages arising therefrom.

Trademark Notice:

SOUL INTELLIGENCE® is a registered trademark of Kristine Genovese. All other trademarks are the property of their respective owners. The use of any trademarks or service marks in this book does not imply any affiliation with or endorsement by the trademark holders.

First Edition

October 2024

For additional resources, information on training, or to connect with the author, visit SoulIntelligenceMethod.com

DEDICATION

To my mom,

You have always believed in me, invested in me, and stood by my side no matter what life brought my way. You are not only my biggest fan, my editor, and my cheerleader, but the ultimate sounding board, my best friend, and my favorite roommate. Your love, support, and wisdom have shaped me, my life, and this journey. This book is as much yours as mine, a testament to the power of belief and the strength of a mother's heart. Thank you for your unwavering support and for always being in my corner.

With all my love,

Kristine

Table of Contents

FOREWORD 1

PREFACE 7

INTRODUCTION 15

CHAPTER 1: UNDERSTANDING SOUL INTELLIGENCE® .21

CHAPTER 2: THE POWER OF EMOTIONS AND ENERGY ..39

CHAPTER 3: CONSCIOUSNESS AND AWARENESS 53

CHAPTER 4: UNPACKING A SOUL INTELLIGENCE® SESSION 71

CHAPTER 5: INTEGRATION AND DAILY PRACTICE 79

CHAPTER 6: BRAINWASHING FOR GOOD 91

CHAPTER 7: THE INTERSECTION WITHHEALTHCARE ..105

CHAPTER 8: THE FUTURE OF SOUL INTELLIGENCE® ...115

CHAPTER 9: TRAINING AND CERTIFICATION 125

CHAPTER 10: PERSONAL HEALING JOURNEYS & TESTIMONIALS 131

CHAPTER 11: THE VISION FOR THE FUTURE 145

CONTACT INFORMATION 152

RESOURCES 153

ABOUT THE AUTHOR 157

FOREWORD

As a medical doctor, I've spent decades grounded in the scientific and evidence-based practices of traditional medicine. I'm board-certified in Bariatrics, Internal, and Integrative Medicine. For years, I studied anatomy, physiology, and biochemistry, dedicated to treating patients with the most advanced techniques in modern healthcare. But I'm also a practitioner who has always believed deeply in the power of holistic healing —an approach that addresses the physical, mental, emotional, and spiritual aspects of health.

Through my work in integrative and functional medicine, I've seen how TRUE healing goes beyond addressing symptoms and what is known in our current traditional medical system. We must dig deeper into the root causes of illness, understanding that the mind, body, and spirit are intricately linked. This philosophy has driven me to constantly search for new and more effective ways to help my patients heal. Decades of research on my medical journey and the many thousands of patients I have worked with for the past 25 years have led me to formulate a complete healing plan for the human body starting at the underlying cause —rather than the symptoms. My program, "The LIT journey" begins with healing the gut, the foundation of our physical health, and culminates with the 11th pillar —spiritual and energetic healing, which I've long believed to be the key to complete well-being.

Despite my commitment to integrative approaches, there were times when even the best treatments seemed to fall short. Some patients would improve physically but remain emotionally or energetically stuck. I knew there was another layer to healing that I hadn't fully accessed. This feeling was amplified by my own personal struggles with stress and burnout. As a mother, doctor, and leader, I found myself depleted, juggling the constant demands of my practice and family, yet unable to restore my health fully. I have lived my life constantly open to growth and commitment to my inner guidance, which has always led me on the correct path. After I struggled deeply with a relapse of fatigue and cardiac issues while under extreme stress, I knew that what it would take to heal was beyond what I was doing prior (even though I was always well beyond the traditional therapies).

That's when Kristine Genovese entered my life. She was one of my patients —an overworked corporate executive with fried adrenals. After receiving treatment for stress-related issues, she reached out to share something UNEXPECTED. Kristine had a gift —a powerful ability to shift energy —and she offered to introduce me to her SOUL INTELLIGENCE® Method, a bioenergetic approach designed to clear the emotional and energetic blockages that can prevent us from healing fully. At first, like many physicians, I was skeptical. After all, I was trained to rely on science and data. But I've always been open to exploring the unknown, especially in the process of healing. So, when Kristine offered to work with me and my staff, I decided to try it.

I remember my first SOUL INTELLIGENCE® session vividly. One of my colleagues had just experienced a profound emotional release. I watched in amazement as she relayed how accurate the information Kristine shared with her was and, more importantly, the clarity and peace she felt afterward. I knew I had to experience it for myself. As Kristine began her work, I felt an immediate connection to the process. It was unlike anything I had ever encountered before—she wasn't just addressing the physical or emotional symptoms; she was tapping into deeper energetic and subconscious patterns that had been affecting my health. By the end of the session, I felt lighter, more aligned, and profoundly clear.

Over the next several weeks, I noticed significant shifts in my energy levels, mood, and overall well-being. My stress levels decreased, and I felt more centered and reconnected to myself. I also saw improvements in my energy and other physical symptoms. The changes I experienced personally opened my eyes to the powerful role that energy plays in health. It wasn't just about balancing hormones or fixing a physical issue—it was about clearing the stuck emotions, deep-seated trauma, and patterns that hold us back.

Inspired by my transformation, I began to refer patients to Kristine. I was amazed by the results. Patients who had been struggling for years with chronic pain, fatigue, anxiety, or emotional trauma were finding relief in ways they hadn't through traditional therapies alone. One patient, who had suffered from migraines for over a decade, saw a dramatic reduction in symptoms after working with Kristine. Another, who had battled depression and anxiety, experienced a deep emotional healing that no medication or therapy had been able to provide.

The beauty of the SOUL INTELLIGENCE® Method is that it complements traditional and integrative approaches to medicine. It doesn't replace them but enhances their effectiveness by addressing the energetic imbalances that often underlie chronic conditions. It also requires zero compliance from patients, they need to show up for a session and receive. I found myself using Kristine's method in combination with functional medicine protocols, and the results were profound. Patients heal faster and more completely. They didn't just feel better physically — they felt more balanced, emotionally stable, and spiritually aligned.

The impact of this method wasn't limited to my patients. Kristine worked with my family, friends, and staff —all of whom had been dealing with their life challenges and the negative impacts of stress. After their sessions, they felt more focused, balanced, and better equipped to handle the challenges of working in a high-stress medical environment. What I've come to realize through my experience with the SOUL INTELLIGENCE® Method is that it's a powerful healing method that addresses the whole person —their mind, body, spirit, and energy field.

Energy is the foundation of our being, and when it becomes blocked or stagnant, it can manifest as physical illness, emotional distress, or a sense of disconnection. By clearing these blockages, Kristine's method allows the body's natural healing mechanisms to take over, supporting a deeper level of healing than traditional or functional medicine alone can provide.

I believe the future of healthcare lies in an integrative approach — one that honors the spiritual and energetic dimensions of health as much as the physical and emotional ones. Kristine Genovese's SOUL INTELLIGENCE® Method is at the forefront of the evolution of medicine, and I'm honored to have witnessed its profound effects firsthand. It has helped to change my understanding of what it takes to heal fully. As we continue to learn about the science behind energy medicine this technique will always be one of my first choices. No doubt that it will continue to transform lives for years to come.

—Dr. Lisa Saff Koche, M.D.

Founder and Director of Spectra Wellness Solutions, and Lit Labs

Triple-board certified physician in Bariatrics, Internal, and Integrative Medicine

Lead Physician and Speaker for Tony Robbins Life Mastery Health Program

PREFACE

The Birth of SOUL INTELLIGENCE® – A Journey from the Corporate World to Spiritual Awakening

A Foundation in the Corporate World My journey into the world of SOUL INTELLIGENCE® didn't begin with crystals, meditation, or anything remotely spiritual. It began in the high-pressure environment of corporate America, where I spent years as a corporate turnaround specialist. My days were filled with strategies, numbers, and a relentless drive to achieve success. I was good at my job—so good, in fact, that I became known for my ability to take struggling companies and guide them back to profitability. It was a role that required precision, intellect, and an unwavering focus on results. I was the embodiment of the corporate warrior, excelling in a world that valued intellect (IQ) above all else.

But beneath the surface of my corporate success, there was a growing sense of discontent. Despite the accolades, promotions, and financial rewards, I couldn't shake the feeling that something was missing. I was fulfilling my professional obligations, but my soul was starved for deeper meaning. The very skills that had made me successful in the corporate world were beginning to feel like a cage, trapping me in a life that was out of alignment with my true self.

The Awakening Begins

The first cracks in my corporate armor appeared as I began to explore personal development and leadership coaching. I had always been driven by a desire to help others, but now I was beginning to see that the traditional corporate methods weren't enough. I started incorporating elements of emotional intelligence (EQ) into my coaching —understanding that how we communicate, empathize, and connect with others is just as important as the intellectual strategies we employ. This was my first step towards a deeper understanding of what it means to be truly aligned in mind, body, and spirit.

However, as I delved deeper into emotional intelligence, I realized that even though it wasn't the complete answer. There was something even more profound that I was missing —something that went beyond the mind and emotions, something that touched the very essence of who we are. I began to sense that there was a deeper level of intelligence that we could tap into, a thing that is connected to our soul, our inner knowing, our divine spark.

This was the beginning of my journey into what I would later call SOUL INTELLIGENCE®.

The Turning Point: Personal Crisis and Spiritual Awakening

As often happens with transformative journeys, my path took a dramatic turn when I encountered a series of personal crises. My marriage, which had been strained for years, finally broke down. I had been living with an under-functioning partner who refused to grow and evolve with me. For years, I had tried to be the caretaker, the one who held everything together, but it was taking a toll on my health. I developed a glaucomic cyclic crisis, where the pressure in my left eye skyrocketed, threatening my vision. My back gave out, and I had to retire from a tennis match for the first time in my life —something that had never happened before. My body was sending me clear signals that something had to change.

In the midst of this physical and emotional turmoil, I experienced a profound spiritual awakening. I began to see that my health issues were not just random occurrences —they were manifestations of the emotional and spiritual dissonance I had been experiencing for years. My body was breaking down under the weight of the unaddressed emotions, unhealed traumas, and unresolved issues that I had been carrying. I realized that I needed to not only heal my body but also address the deeper root causes that were at the heart of my suffering. This awakening led me to explore various spiritual practices, energy healing modalities, and alternative therapies. I was particularly drawn to the idea that our bodies are not just physical entities but are deeply interconnected with our emotions, thoughts, and spiritual energies. I began to experiment with different techniques, including meditation, visualization, and energy work, all to heal myself on a holistic level.

The Birth of SOUL INTELLIGENCE®

As I delved deeper into these spiritual practices, I started noticing something extraordinary. When I combined the intellectual rigor of my corporate background with the emotional awareness of EQ and the spiritual insights I was gaining, I started to see profound shifts in myself and those I worked with.

It was as if I had unlocked a hidden dimension of intelligence —one that went beyond the mind and emotions and tapped into the very essence of the soul.

This realization was the birth of SOUL INTELLIGENCE®. Through meditation and the creation process, it felt like I was remembering ancient wisdom and tapping back into a gift that had been given to me. I began to develop a method that could help people access this deeper level of intelligence, one that could guide them to heal not just their bodies and minds but their very souls. I saw that when we align our mind, body, and soul, we become truly powerful beings, capable of creating profound change in our lives and in the world around us.

SOUL INTELLIGENCE® is about tapping into that inner knowing, that divine spark that resides within each of us. It's about clearing away the old beliefs, patterns, and traumas that have been holding us back so that we can align with our true mission, vision, values, and purpose. It's about living from a place of authenticity, where we are fully in alignment with our higher selves and the divine purpose for which we were created.

The Science and Spirituality of SOUL INTELLIGENCE®

One of the most fascinating aspects of SOUL INTELLIGENCE® is the way it bridges the gap between science and spirituality. On the one hand, it draws on the latest research in quantum physics, bioenergetics, and neurobiology —fields that are beginning to uncover the profound ways in which our thoughts, emotions, and energies shape our physical reality. On the other hand, it also taps into ancient spiritual wisdom, recognizing that we are spiritual beings having a human experience and that true healing comes from aligning all aspects of our being —mind, body, and soul.

At the heart of SOUL INTELLIGENCE® is the understanding that we are all connected by a vast, universal energy field. This field, sometimes referred to as the quantum field or the morphogenic field, is the source of all creation. It's the place where our thoughts, emotions, and intentions interact with the fabric of reality to bring about change. When we learn to align our energies with this field, we can create profound shifts in our lives.

One of the key principles of SOUL INTELLIGENCE® is that emotions are energy in motion. When we experience negative emotions —such as fear, anger, or sadness —these energies can become stuck in our bodies, leading to physical, emotional, and spiritual disease.

Imagine your body as a river, with energy flowing like water. When the river is clear, water flows smoothly and easily, nourishing the banks of the land. But when debris like negative emotions or unresolved trauma —blocks the river, the water stagnates, pools, and overflows. The SOUL INTELLIGENCE® method helps you remove that debris, allowing your energy to flow freely, and allowing the body's natural healing processes to take over and revitalize your entire being.

The Method in Action: Healing at The Soul Level

SOUL INTELLIGENCE® is a bioenergetic healing system that works by aligning the mind, body, and soul. It's a process that recognizes the importance of addressing all aspects of a person's being —not just the physical symptoms or surface-level emotion but the deep-seated beliefs and energetic patterns that underlie them. The method involves a combination of meditation, visualization, and energy work, guided by a series of detailed charts that help to identify the specific issues that need to be addressed. Using a pendulum, I tap into the divine guidance that flows through me, identifying the energies, emotions, and patterns that needed be clear for the individual's highest good. Once these blockages are identified, we work together to clear them, creating space for new, supportive energies to flow.

The beauty of the SOUL INTELLIGENCE® method is that it doesn't require the individual to relive past traumas or engage in lengthy therapy sessions. Instead, it works at the energetic level, clearing the blockages and allowing the individual to move forward with greater ease and alignment.

From Corporate Success to Spiritual Fulfillment

As I began to integrate SOUL INTELLIGENCE® into my coaching and consulting work, I saw incredible transformations in my clients. CEOs, healthcare professionals, and individuals from all walks of life began to experience profound shifts —not just in their professional lives but in their personal lives as well. They became more aligned with their true purpose, more connected to their inner wisdom, and more empowered to create the life they truly desired.

For me, this journey has been nothing short of miraculous. What began as a personal quest for healing and fulfillment has grown into a movement —one that I believe has the potential to change the world. SOUL INTELLIGENCE® is not just a method; it's a way of life, a way of being that allows us to live in alignment with our highest selves and our divine purpose.

In the following Chapters, I will share more about the science and spirituality behind SOUL INTELLIGENCE®, the method itself, and the incredible transformations that have come from it. I'll also delve into the practical applications of SOUL INTELLIGENCE® —how it is utilized in business, healthcare, and personal development to create profound and lasting change.

I hope this book will serve as a guide for anyone who is seeking to live a more authentic, aligned, and fulfilling life. Whether you are a leader looking to inspire and motivate your team, a healthcare professional seeking to improve patient outcomes, or simply someone on a personal journey of healing and growth, I believe that SOUL INTELLIGENCE® has something to offer you.

So, join me on this journey as we explore the power of SOUL INTELLIGENCE® and discover how it can transform your life, your work, and your world.

INTRODUCTION

Healing Through Soul Intelligence®
How To Use This Book

Welcome to Healing Through Soul Intelligence®. Whether you're new to energy healing or an experienced practitioner looking to deepen your understanding, this book is your guide to unlocking the profound potential of your own Soul Intelligence® —the innate wisdom that resides within each of us. Throughout the following chapters, you'll learn not only what Soul Intelligence® is but also how to use it to heal yourself, enhance your personal growth, and help others.

This book is designed to take you on a journey of self-discovery, healing, and transformation. The SOUL INTELLIGENCE® Method taps into the power of bioenergetics to help you access and release trapped emotions, resolve trauma, and shift self-limiting beliefs. The result? A clearer, more aligned version of you —one that's deeply connected to your higher self and your soul's purpose.

How To Get the Most Out of This Book

This is not just a book you read once and set aside. It's a resource you'll want to return to again and again as you move forward on your healing journey.

To help you maximize your experience, here are a few suggestions on how to best use this book:

1. Take Your Time

Each chapter of this book contains layers of information, personal stories, and practical exercises designed to help you connect with your Soul Intelligence®. Don't rush through it. Instead, give yourself time to digest each section, reflect on its meaning, and apply what resonates with you. Healing is a process, not a race. The insights shared here are meant to be integrated into your life gradually.

2. Start With an Open Mind and a Heart

One of the key tenets of the SOUL INTELLIGENCE® Method is the importance of an open mind and heart. Whether you're familiar with bioenergetics and energy healing or this is your first exposure, approaching the material with curiosity and openness will allow you to fully receive the healing potential of this work. Let go of any preconceived notions about how healing should look or feel and be open to new possibilities.

3. Journal Your Journey

Healing through Soul Intelligence® is an ongoing practice. As you move through this book, it's helpful to keep a journal where you can document your thoughts, feelings, and experiences. Writing helps you process what's coming up for you emotionally, mentally, and spiritually and allows you to track your progress over time. Use the journal prompts provided throughout the book as starting points for your reflections.

4. Practice The Techniques Regularly

This book is filled with practical techniques —meditations, visualization exercises, and the Sacred Soul Script journaling practice —that are designed to help you connect more deeply with your own energy field and higher self. Don't just read about them —practice them. The more you engage with these techniques, the more effective they become. Set aside time each day or week to incorporate these practices into your routine.

5. Reflect And Revisit

Healing happens in layers. You may find that certain chapters or exercises resonate with you at different times in your life. As you evolve, so too will your understanding of the material. Don't hesitate to return to earlier chapters or re-read sections that feel particularly relevant. Each time you revisit a concept or technique, you'll uncover new insights and deeper levels of understanding.

6. Apply Soul Intelligence® To All Areas of Your Life

The principles of Soul Intelligence® can be applied to all aspects of your life —your relationships, career, health, and personal growth. As you progress through the book, you'll learn how to access your Soul Intelligence® to guide you in making decisions, releasing old patterns, and aligning more fully with your highest purpose. Don't limit this work to one area of your life —use it to create shifts in every dimension of your existence.

7. Engage With Group Healing and Collective Energy

One of the most powerful aspects of Soul Intelligence® is its ability to affect not just individuals, but entire groups and collective energy fields. As you move deeper into your own healing, consider participating in group clearings or working with others to create positive shifts in shared spaces —whether it's your family, workplace, or community. You'll find guidance in the book on how to engage with this collective energy.

8.Honor Your Own Timing

Healing is a deeply personal journey, and it doesn't follow a linear path. Some chapters or exercises may resonate with you immediately, while others might take time to fully understand. Trust your own process. It's okay to pause, reflect, or move more slowly through certain sections. Your healing will unfold exactly as it's meant to.

9.Incorporate The Spiritual and Energetic into Your Daily Life

Healing through Soul Intelligence® goes beyond isolated practices or occasional moments of insight. It's about living in alignment with your soul's purpose on a daily basis. As you work through this book, look for ways to bring these teachings into your everyday life — whether that's through morning meditations, daily affirmations, or simply becoming more mindful of your energy and emotional state throughout the day.

10.Share Your Journey

Healing is a deeply personal experience, but it's also one that can inspire and uplift others. As you progress on your own path, consider sharing your experiences with friends, family, or within a community of like-minded individuals. Your journey may encourage others to explore their own Soul Intelligence® and start their own healing process. The more we share this work, the more we raise the collective vibration and contribute to a healthier, more harmonious world.

What You'll Learn

In this book, you'll find a wealth of information on the SOUL INTELLIGENCE® Method and how it can be applied to various aspects of your life. Some of the key topics we'll cover include:

Understanding Soul Intelligence®:

What it is, how it works, and why it's such a powerful tool for healing.

Bioenergetics And Healing:

How your energy field influences your physical, emotional, and spiritual health.

Consciousness And Awareness:

The levels of consciousness we can access and how they shape our experience.

Clearing Trapped Emotions and Trauma:

How to identify and release the emotions and trauma that are holding you back.

Integrating Soul Intelligence®:

Practical tools for incorporating these teachings into your daily life, from meditation to journaling.

Group Healing and Collective Energy:

The power of working with collective energy fields to create shifts on a larger scale.

The Future of Healing:

How Soul Intelligence® can transform healthcare, business, and personal development.

This book is more than just a guide —it's an invitation to step into a new way of being. As you engage with these teachings, you'll begin to experience the freedom, clarity, and empowerment that comes from aligning with your Soul Intelligence®. It is my hope that this book will not only help you heal, but also inspire you to live a life of purpose, authenticity, and deep connection with your true self.

Let the journey begin.

Kristine Genovese

CHAPTER 1

UNDERSTANDING SOUL INTELLIGENCE®

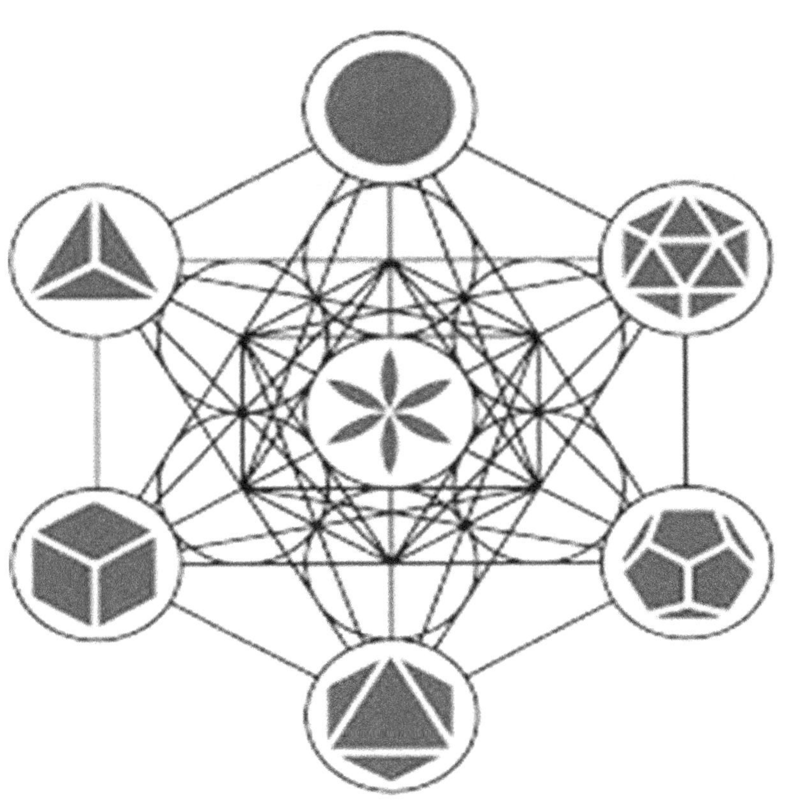

What is Soul Intelligence®?

Imagine that you're a captain at the helm of a magnificent ship. You've charted your course toward your destination and have the latest navigation technology (that's your IQ), and you've mastered the art of reading the waters and the skies (that's your EQ). But there's a hidden force that goes beyond maps and instruments —a deep, magnetic pull that guides you, even when at times you may feel lost at sea. This is your SOUL INTELLIGENCE® (your SQ).

SOUL INTELLIGENCE® is like the compass that always points you toward your true north. It's innate knowing that, no matter the storm, there's a path beneath the waves that always helps you navigate. When you tap into your SOUL INTELLIGENCE®, you're not just a captain steering with skill; you're guided by the stars, the tides, and the vast energy of the ocean itself. It's the magic of being connected to something much greater than yourself.

Imagine now, as you sail, you feel the ship's vibrations, hear the creak of the wood, and sense the rhythm of the sea. Each sway and surge are in tune with the primordial sound of the universe, the OM resonating a divine energy in every wave. This is Soul Resonance —when every part of you vibrates in harmony with the world around you.

Now, picture the crew: they're not just following orders; they're driven by a shared purpose. Each one of them is alive with passion, not just for the journey but because they're aligned with the ship's mission. That's the power of SOUL INTELLIGENCE®—when you align your mind, heart, and spirit, you become unstoppable.

It's the secret ingredient that turns an ordinary voyage into an epic adventure.

Or think of it like this: Imagine you're a musician. You've learned all the notes, mastered the scales, and can read the sheet music perfectly. But there comes a moment when you let go of the how and simply feel the music flowing through you. You're no longer just playing; you're creating something new, alive, and magnetic. The music becomes part of you, and you become part of it. That's Soul INTELLIGENCE®. It's when you stop thinking and feeling and start being, knowing, and flowing. It's when you're in such sync with the rhythm of life that your actions, thoughts, and emotions are like notes in a song, effortlessly blending with the world around you.

SOUL INTELLIGENCE® isn't just about solving problems or achieving goals. It's about living fully, with an unshakeable sense of purpose, tapped into a wisdom beyond logic and an energy beyond emotion. It's about realizing that, like the captain guided by the stars or the musician lost in the melody, you are in harmony with a greater intrinsic force—one that not only lights your path but also lights you up from the inside out. That's living, loving, and leading your life by the power of SOUL INTELLIGENCE®.

The Birth of Soul Intelligence®

The concept of SOUL INTELLIGENCE® didn't come to me through a single flash of insight; rather, it was the culmination of years of exploring the interplay between intellect, emotion, and spirit. My journey into SOUL INTELLIGENCE® began in the corporate world, where I had mastered the power of intellect (IQ) to turn around struggling companies. Then, as I began incorporating emotional intelligence (EQ) into my leadership coaching, I realized I was still scratching the surface of something far deeper. There was an inner knowing, a spark within us all that connected to something greater. This was the realm of SOUL INTELLIGENCE®—a place where intellect, emotion, and the soul converge.

SOUL INTELLIGENCE® is about operating from a place of deep alignment with your mission, vision, values, and purpose. It's tapping into the divine essence within you, your higher self, and allowing that part of you to lead your life. It's about navigating all of life's experiences from that space of inner knowing, where you embody the best version of yourself every single day.

That's what it means to live in alignment with Soul Intelligence®. It's why I often say it is literally the answer to all of your problems. Each of us truly has all of the answers to all of our challenges. In essence, it's because what's going on inside of us manifests outside in the material world. This phenomenon can also be explained as soul resonance—the harmonic vibration between your inner being and outer world. When you achieve this state, you experience synchronicities, effortless flow, and a profound sense of peace. SOUL INTELLIGENCE® provides the tools and insights needed to cultivate this resonance.

Core Principles of Soul Intelligence®

We all have a surface understanding of IQ and EQ, but SOUL INTELLIGENCE® (SQ) is the next step in human evolution. It's about knowing, at a soul level, that you already have the answers to your greatest challenges. When we step back from the minutiae of daily life, we gain a higher perspective—a 30,000-foot view—allowing us to access deeper awareness and clarity. The key to SOUL INTELLIGENCE® is learning to trust that inner voice, tuning into your body, and listening to what it's telling you.

When I began incorporating the concept of SOUL INTELLIGENCE® into my coaching practice, I needed to be able to speak in plain language to CEOs of companies. I asked them to imagine how their company would function if every single-person organization was lit up from the inside out. What if your employees were totally motivated, inspired, excited, and in alignment with your mission, vision, values, and purpose? What if they also felt like being a part of your company allowed them to be in alignment with their own personal mission, vision, values, and purpose? Can you envision how productive your people would be? What achievements could your company experience? Your people would run through the fire for you because they are lit up from the inside out and highly motivated.

I then went on to explain that you understand the concept of IQ. Intelligence, intellect, ideas. We know we need smart people who can figure, strategize, invent, and think. In the 80s and 90s, the concept of EQ—emotional intelligence—came into the workplace. Knowing "how to say" what you wanted to say became more important than what you were actually saying.

Understanding emotions and providing the "why behind the what" was now an expectation in the workplace, not just a nice thing to do.

But neither of these things can completely tap into someone's intrinsic motivation.

So, I would ask, have you heard of SQ? SOUL INTELLIGENCE®? That's someone's inner knowing, their divine talent, their magnetic spark, their zone of genius. If you can get someone's mind, body, and soul to be in alignment—that's the magic we all want to experience in life. I ask you; what CEO would turn me down? The answer is none. I had the most successful year of my career, earning more that first year on my own than I had in my previous year at the "big corporate gig." The Divine blessed me for following my true path.

Our bodies serve as powerful barometers, constantly signaling our emotional state, beliefs, traumas, and unresolved issues. The energy of our emotions is stored in our cells, and unless we address this energy—feel into it, release it—it can block us from reaching our highest potential. SOUL INTELLIGENCE® helps us tap into this inner wisdom, enabling healing and transformation at the deepest levels.

What I help people discover is how to access and operate from their SOUL INTELLIGENCE®, a place of inner knowing. That's when you know you are fully in alignment with your mission, your vision, your values, and your purpose, and you feel like you're part of something bigger than yourself.

That's being in touch with that eternal part of you, that divine spark within you, your Higher Self, the best version of you. How do you allow yourself to be led from that place? How do you live your life from that place? How do you navigate all of your experiences from that place?

How do you embody the best version of yourself daily? That's operating from your SOUL INTELLIGENCE®.

SOUL INTELLIGENCE® is comprised of four pillars:

1. Conscious Awareness – Understanding the current state of your mind, body, and soul.

2. Energetic Clearing – Using the method and other techniques to remove blockages and toxic energy.

3. Spiritual Integration – Methods to align your life purpose with your spiritual path.

4. Soul Expansion – Practices to continuously grow and evolve on a soul level.

Throughout this book, we will explore how to master each pillar.]

The Essence of Soul Intelligence®

My own journey through the higher education and functional medicine worlds introduced me to a diverse range of integrated healing modalities. From red light therapy to Reiki, from cryotherapy to Theta Healing, I saw the potential in treating the whole person—mind, body, and spirit. Yet something was still missing. While these methods were powerful, I realized that SOUL INTELLIGENCE® offered something more—an ability to bring what is buried in the subconscious to the surface in a safe, directed, and specific way. It's a bit like Reiki, but more intentional, with a focus on releasing subconscious blockages that are holding us back from alignment.

Neville Goddard captured the essence of this work when he said, "Emotion is the secret." You can think all the right thoughts and say all the affirmations you want, but if your emotions are not aligned with those thoughts, you won't manifest your desires. Your subconscious responds to emotion and belief. Emotion is what amplifies and magnetizes your thoughts into reality. It's the missing piece in many personal development practices, and SOUL INTELLIGENCE® provides the tools to bring that emotion into alignment with your highest self.

Energy healing, once relegated to the fringes of spirituality, is now gaining recognition in the scientific community through the emerging field of biofield science. Researchers are studying the biofield, the complex energy field that surrounds and permeates the human body, and its role in health and disease. Techniques like Reiki and acupuncture are believed to interact with this biofield, promoting balance and healing. This scientific exploration validates the efficacy of energy healing practices, offering a bridge between ancient wisdom and modern medicine. Many registered nurses are being trained in Reiki to help patients heal more quickly. What was once thought of as "woo-woo" is now being used in hospitals as part of patient care plans.

The Science Behind Soul Intelligence®

Albert Einstein once said, "Everything is energy, and that's all there is to it. Match the frequency of the reality you want, and you cannot help but get that reality. It can be no other way. This is not philosophy; this is physics."

Let's talk about energy. So, how does this all work? Think about it. We all know from 4th grade science class that everyone and everything is energy—protons, neutrons, electrons, vibration, electromagnetic field… That means that energy can get stuck in people, places, or things—AND in your mental, emotional, spiritual, or physical body.

Everything is energy, vibration. Everything, from our thoughts to our emotions, carries a frequency. Emotions like fear, anger, and guilt vibrate at lower frequencies, while love, joy, and peace resonate at higher frequencies. Our states of consciousness also align with different brainwave frequencies—alpha, beta, theta, delta, and gamma—and our physical and emotional well-being is deeply connected to these energetic states. Dr. David Hawkins actually measured these states of consciousness and corresponding frequencies in his groundbreaking work Power vs. Force.

We all spend a lot of time in our heads, so let's start at the top with our minds. Did you know that we only use 5% of our mind to do ALL of our thinking—all the figuring, strategizing, factoring, calculating…5%! That's it! Crazy. 95% of our brain activity is subconscious! That means negative energy, toxic emotions, destructive thoughts, repetitive caustic patterns, stored trauma, and self-limiting beliefs reside in your subconscious mind and typically loop—cycling over and over again. Our subconscious mind is in the driver's seat, and it's powerful. It only pays attention to the thoughts around the feelings—when you feel, when you believe you deserve something, you get it.

Positive or negative. It's how the quantum field of the universe works. So how are you supposed to know what you need to shift if you don't even know how to access it?! That's where the power of SOUL INTELLIGENCE® comes in!

In my SOUL INTELLIGENCE® practice, I use techniques that help clients shift their energy, bringing the subconscious into the conscious mind for healing and release. The energy that gets stuck in the mind, body, and spirit must be cleared for true healing to occur. It helps you clear away old beliefs that are no longer serving you, old patterns or archetypes that aren't supporting you, stuck trauma, and feelings that have been suppressed—so you can more easily go within and become aligned to your true calling, your mission, your vision, your values, and your purpose—your why. This incredible methodology allows you to access that information wherever it's residing without having to relive any trauma around it.

Imagine breaking free from the chains of past traumas that have silently dictated your life choices, relationships, and self-worth. SOUL INTELLIGENCE® provides you with tools to help you transform, feel empowered, and reclaim your life. This method has the potential to restore your sense of self, bring peace to your mind, and rekindle the joy you may have lost. This is not just about coping;

it's about thriving. This bioenergetic system works on a quantum level, tapping into the universal energy field to create profound change.

Soul Intelligence® as a Quantum Healing System

SOUL INTELLIGENCE® operates on the cutting edge of science and spirituality. Quantum physics tells us that we are all connected through a universal energy field, and the SOUL INTELLIGENCE® method taps into this field to create healing at the deepest levels. It works by aligning the mind, body, and soul, clearing away old beliefs, patterns, and traumas that no longer serve us.

One of the most remarkable scientific findings in recent years has come from the HeartMath Institute, which discovered that the heart has an electromagnetic field six times larger than the brain's. In fact, the heart is not just some random pump in your chest—it actually has a six-foot-wide electromagnetic field around it, whereas the brain merely has a two-foot-wide field. Which is pretty incredible when you consider how much time we spend thinking, strategizing, and using our brain.

This electromagnetic field responds to our emotions, and when our heart and mind are in coherence, we create a state of alignment that attracts what we truly desire. The SOUL INTELLIGENCE® method leverages this understanding, helping individuals align their emotions and intentions for powerful transformation.

The quantum field is an electromagnetic field, and it all starts with the electric spark—that masculine energy with IQ—intellect, ideas.

But what magnifies and brings it into form is the magnetic quality of the feminine energy—EQ Emotional Intelligence, emotions. The concept of emotional intelligence is deeply rooted in both spiritual wisdom and scientific research. The HeartMath Institute has discovered that the heart has its own "brain," a complex nervous system that processes information independently of the brain in the head. This heart-brain connection is key to understanding how our emotions influence our decisions and overall well-being. By nurturing emotional intelligence, we harmonize the heart and mind, aligning with both spiritual teachings and cutting-edge science to foster personal growth.

Emotions are the magnetic force in the electromagnetic field, the amplifier that collapses the particle wave and brings thoughts into form. That's why it's not enough to just "think positive"—you must feel the emotions that match what you're trying to create. When we align our emotions with our intentions, we tap into the true power of the quantum field.

Let's go a bit deeper into the science of energy by exploring our bodies biology as to how and why energy gets trapped within our organs and systems. Our bodies are protein-making machines. Every cell in our body except red blood cells makes proteins. And, in order for a cell to make proteins, a cell has to be stimulated or regulated.

Our genes are essentially a library of potentials. And the latest research in the field of genetics says genes don't create disease; it's the environment that signals the gene that begins to create disease.

In fact, less than 1% of people are born with genetic conditions—the other 99% is created from lifestyle, behavior, and choices.

We can talk about lots of external environmental factors that affect our health. For now, let's focus on our inner environment, as it truly is the correlation between our thoughts, emotions, and our physical body. Our inner environment is affected by our emotional state, which is a key factor in sending signals that cause our genes to upregulate or downregulate. The inner environment signals the gene, and the end product of an experience is called an E-motion (energy in motion).

So, when someone has an emotionally charged experience, it begins to send a chemical-emotional signal to their body. And it's that chemical emotional signal that begins to change the person's state of being.

The person's body and brain are now effectively operating in the past as that event has changed them biologically.

What happens to most people is that they begin to create beliefs from their past experiences. For example, any kind of memorable trauma, like a car accident or physical abuse. The trauma stays with you; you form a belief like "I can't win," and you begin to operate in the world like looking through a dirty shower door. You tend to look for more evidence in every situation to prove or validate your belief. Sometimes the beliefs form into a repetitive pattern from childhood—perhaps you were told you weren't "good enough" as a kid, so as an adult you became an overachiever like yours truly.

We are our thoughts neurologically and our feelings chemically. Our bodies become the reflection of our internal condition. When we embrace the change in our biology as a result of our past experiences, we have a new state of being. In short, how we think and how we feel creates our state of being. Highly charged emotional events can signal genes that begin to cause the body to change. Further, most people live in a routine—internally and externally. A licensed psychologist, Wallace K. Pond, Ph.D., said, "Some people prefer the certainty of misery over the misery of uncertainty." He was referring to the reason why people stay stuck in toxic relationships or dead-end jobs.

If people are living in the routine of the predictable future, or their memories are associated with the emotions of past experiences or trauma, that routine has begun to influence their gene expression. Why? Because there is no new information coming in from the environment. The repetitive negative thoughts in their thinking lead to self-limiting beliefs; the toxic emotions that they continue to feel lead to repetitive patterns—constantly sending the same information to the genes over and over again. (Remember the definition of insanity is doing the same thing over and over again and expecting different results.)

The genes in your body need a pattern disruptor or you will repeat the same old, same old.

Once the gene has been signaled and the person is constantly living out of balance, that imbalance becomes their new normal. The internal environment of the body is sending specific signals to the cells, and those cells are getting the same instructions. The same instructions produce the same proteins. The person begins to experience ill health or dis-ease within the body, or a dissonance in the frequency, and they have to go to a healthcare professional.

Hello chronic illness, hello repetitive conditions, hello groundhog patients, hello fix one malady, then another one surfaces—sound familiar? It's also why depression is so pervasive. The same information to the cells, repeating the same condition. When you have an internal experience in your mind that carries an amplitude of energy that is greater than the hardwired programs in your brain and the emotional conditioning in your body—then you will become biologically changed for better or for worse. It's your choice how you react to the experience—positively or negatively. Emotional states change the internal mapping of your brain.

Your biology literally changes, or better yet, the past no longer exists. In essence, you rewrite a new pattern; new information comes into the body. Energy influencing matter is a quantum phenomenon.

We all have the ability to be quantum healers. SOUL INTELLIGENCE operates on a bioenergetic level—a quantum form of healing!

Even Albert Einstein said the future of medicine was frequency! He is credited with saying, "Future medicine will be the medicine of frequencies." Frequency medicine, also known as vibrational medicine, uses vibrational energy to restore health and vitality. It can be used to diagnose and treat illnesses by measuring the frequencies of the human body and its electromagnetic field.

Not only can changing your internal state help you heal physically and make you feel and operate mentally and emotionally better—it could actually help you rewind the clock on aging. One of the pioneers in neuroscience, Dr. Joe Dispenza, shared the following experiments supporting this epigenetic research: people with type 2 diabetes were exposed to a very funny one-hour comedy show, where they laughed hard for an hour. The experiment measured the before and after to see whether or not their genes were upregulated. The results were astounding! Twenty genes were upregulated just by people changing their mood! So, by changing their internal state, they were able to instruct new genes to upregulate and make new healthy proteins!

Studies out of UCLA have shown that men with prostate cancer were able to upregulate new genes for health and downregulate those genes for disease by meditating daily for 60 days. Not only that, but they also lengthened their telomeres and increased their longevity (their biological age vs. their chronological age).

Dr. Joe Dispenza regularly holds four-day meditation retreats. People who attended were able to reverse serious health conditions in just four days! Let's get Zen, my friends! When people begin to heal their genes, they are up-regulating new genes and down-regulating other genes that create disease or imbalance.

Proving that we are our own genetic engineers and not our genes.

As we align our energy with our higher purpose, we become powerful creators of our reality. Our thoughts, emotions, and intentions work together to shape our experiences, and SOUL

INTELLIGENCE® provides the roadmap for this alignment. Whether you are healing from past trauma, seeking clarity on your life's mission, or simply looking to live more authentically, SOUL INTELLIGENCE® offers a pathway to transformation.

In the chapters that follow, I will dive deeper into the science and spirituality behind SOUL INTELLIGENCE®, offering practical tools and insights to help you live from a place of alignment and inner knowing.

Together, we will explore how this powerful method can help you manifest your highest potential and create profound change in your life and the world around you.

Journal Prompts

How do you embody the best version of yourself daily?

What surfaces for you when you read this quote: "Some peopleprefer the certainty of misery over the misery of uncertainty."

CHAPTER 2

THE POWER OF EMOTIONS AND ENERGY

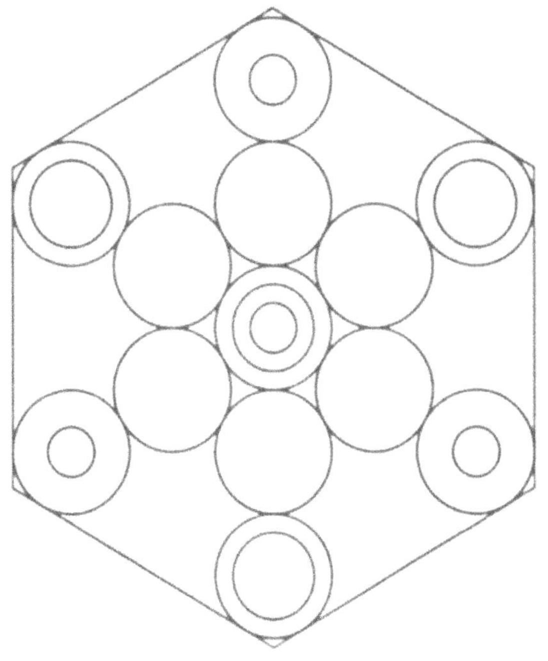

Emotions: The Secret to Healing

In the world of SOUL INTELLIGENCE®, emotions are not just fleeting feelings; they are powerful currents of energy that shape our reality. Emotions are, quite literally, energy in motion. When this energy flows freely, we experience health and harmony. However, when emotions are suppressed, the energy becomes trapped within our cells, creating blockages that can lead to dysfunction and disease in the body. We often suppress our feelings, and when we do that for an extended period, that's when you start to experience dysfunction, a dissonance in your frequency, and ultimately disease within the body. Our bodies, ever attuned to these energetic disturbances, often manifest these blockages as physical discomfort or chronic illness.

Emotional healing is often seen as a simple process of "letting go," but SOUL INTELLIGENCE® goes far beyond that, offering a profound exploration of the intricate layers involved in true emotional recovery. Understanding what situations are still being held within your system, when they occurred, and what energies are still held in place—it's extremely empowering. It delves into the subconscious mind where our deepest wounds are stored and examines how these unresolved emotions manifest in our physical health, relationships, and overall well-being. It allows us to dive into the nature of the emotional pain and the liberation that comes from healing it at its source. Clients always come away with some new insight, a level of self-awareness, and a deepened capacity for self-compassion and healing.

I can remember experiencing the power of emotions and energy very early on in life. I recall telling my mom, "I don't like how I feel when I'm around my dad because I feel like I change physically." What I recognized as an adult is that I was literally taking on his pain, self-loathing, grief, jealousy, and rage. I was literally absorbing his emotions and would wind up feeling exactly how he felt about himself. I was an energetic sponge, unconsciously absorbing his toxic emotions to try and help heal him. Working with energy requires a new level of awareness as to how not to take on other people's trauma; otherwise, you may experience some physical discomfort. My body lets me know when I've done this initially by bloating my belly, then gastrointestinal tract distress ensues. I know it's time to get out my book and pendulum and shift some things up and out so that they're transmuted back to the Divine Source as opposed to staying in my energy field.

It's also important to note that when you are in close proximity to someone or exchanging body fluids of any kind, you're exchanging energy. Whether you are kissing someone, holding hands, being intimate, or even next to someone—you are literally exchanging energy. If you have empathy for them, you may also have a propensity to take on their essence, their emotions, their struggles, or even their grief. Oftentimes, when you are physically ill, it's not just your own toxic emotions causing it but those around you.

The root cause of all physical discomfort and potentially chronic illness is trapped emotion. This isn't just a theory; it's reflected in staggering statistics. According to the American Council on Aging, around 60% of adults suffer from chronic illness, and 40% live with at least two chronic diseases.

The SOUL INTELLIGENCE® method gets to the heart of these issues by addressing the emotional energy that lies at the root of our physical and mental ailments.

By releasing this trapped energy, we can initiate profound healing.

Quantum Physics and the Soul Intelligence® Method

The SOUL INTELLIGENCE® method is grounded in the principles of quantum physics, which reveal the profound interconnectedness of all things. Think of the quantum field as an invisible web connecting everything in the universe. Just like pulling on one strand of a spider's web can cause vibrations across the entire structure, changing your energy can create ripples throughout your life and beyond.

Recent advancements in quantum physics suggest that consciousness is not just an abstract concept but a measurable force that interacts with the very fabric of reality. The double-slit experiment, for example, demonstrates how observation (a function of consciousness) can influence the behavior of particles. This revelation suggests that our thoughts and intentions have a direct impact on the material world, merging the boundaries between spirituality and science in a profound way.

Quantum physics teaches us that everything is energy, vibrating at different frequencies. This includes our thoughts, emotions, and even our physical bodies. When we understand that we are beings of energy, we begin to see how our emotions, which are also energy, can influence our physical reality. Instead of thinking of your energy as something abstract, imagine it as a battery that powers your entire life. When your battery is low, you feel tired, stressed, and unmotivated. SOUL INTELLIGENCE® helps recharge that battery, restoring your energy so you can live fully and freely.

The SOUL INTELLIGENCE® method leverages this understanding by helping individuals release the trapped emotional energy that causes dissonance in their frequency. By bringing these energies from the subconscious to the conscious mind, we can clear them, allowing for a restoration of balance and well-being. This approach aligns with the latest research in bioenergetics, which studies the relationship between our energy fields and our physical health.

Interconnectedness of Mind, Body, and Soul

In a world that often compartmentalizes the mind, body, and spirit, SOUL INTELLIGENCE® offers a deeply integrative perspective, revealing how these aspects of our being are inextricably linked. Our thoughts share our physical health, our emotions influence our mental clarity, and our spiritual well-being affects our overall vitality. In my own life, I've seen firsthand how deeply interconnected our mental, emotional, physical, and spiritual health are. In 2023, I was under immense stress. I was an over-functioning woman married to an under-functioning man who had long since stopped evolving with me.

Despite my pleas, prayers, and efforts to encourage him to grow and heal, he remained stuck, focused on others' successes while neglecting his own. He had quit on himself, let alone us—years prior to when I actually had the courage to leave. I had prayed, tolerated, pleaded, endured, and begged him to get help, to change. I wanted him to wake up and grow with me.

In fact, at that time, I was a life coach. I helped to guide others to grow and assist them in achieving their goals. How could my own husband not want to live his best life? One of my gifts is to see the best version of someone—their higher self. You know, that wise one who is so proud of how far you've come and is so excited about the journey that still lies ahead. I actually work to shift energy up and out of people that's no longer serving them. Yet I couldn't help him.

Everyone has free will. Unfortunately, his choice was to stay stuck, depressed, focused on others' success, and not motivated to create his own. My choice was to continue to grow, evolve, and help others while becoming the best version of myself every day, all the time. The gap between us finally became too great.

The crazy thing is that I didn't recognize it until I physically broke down. I nearly blinded myself because I didn't want to see our life the way it was and felt helpless to change it. I was seeing him constantly suffering and unhappy. I was always trying to make him happy and felt hopeless and defeated because he wouldn't change. Further, I didn't feel like I was a priority for him. In fact, none of my needs were being met. I felt invisible.

My own body began to bear the weight of this emotional burden. The pressure I was under manifested physically. I developed a glaucomic cyclic crisis, with the pressure in my left eye reaching dangerous levels. Glaucoma is pressure on the optic nerve that causes blindness. My back gave out, leaving me unable to participate in my one healthy outlet—tennis. I went on the court and had to retire after three games because I couldn't walk. In the 20+ years of playing tennis—I never had to retire from a match! My hip had gone out of joint, which represents feeling used, unappreciated, let down, unsupported, taken advantage of, unacknowledged, and holding too much responsibility. These physical symptoms were my body's way of screaming for attention, telling me that enough was enough.

Even still, I stayed. All the while, he was focused on his own health issues. Prostate cancer, depression, then a heart event landing him in the hospital. It was at that moment that I had my breaking point. I was falling apart, and he was in no condition to help take care of me. It was again all about him.

His illness, his trauma, his anxiety, his depression, his life. He refused to see that we create our own life circumstances that support our beliefs. He refused to take care of himself. He didn't support his body with nutrition or the supplements necessary to help support his immune system. In fact, he continued to drink alcohol and eat poorly, which actually fed the cancer. He was self-sabotaging, and I was helpless to change it.

That's probably the hardest thing to accept in this life. You cannot change anyone else. You can only change you. They need to want to heal. They need to want to grow. They need to internally create the life they want and work towards their goals. My life was so out of balance that the only thing I could do was to leave to create a peaceful environment for me to heal. So on July 4th, I left. I left my old life behind and decided it was time to focus on the life I've been dreaming about. The life that I journal about. The life I meditated about. The life I'm meant to lead.

The discomfort in our bodies doesn't have to be that dramatic for us to notice that something is off. You might just be having a challenging week where you feel tense and aggravated, you start to experience neck pain, so you choose to get a massage. The root cause is typically a romantic partner is a pain in your neck, so you have to get underneath that and explore what are they mirroring for you to shift up and out?

When this happens to me, I use my SOUL INTELLIGENCE® method to lessen some of the hold the energy has on me first. When you've held the energy in the body long enough, sometimes you have to have physical manipulation. In order for this to completely heal, we can attack it energetically. You can also work on it through meditation and prayer, and by sending healing energy into the body, but if it's been in there long enough, you might have to go to a chiropractor and perhaps get massage therapy to manipulate the rest of it out of the physical because you let it stay there so long.

Again, this is what your body is designed to do. It's really to be this indicator of what's going on emotionally and energetically. What are you processing? What are you holding onto? How is this shaping your experience? You have a physical body, a mental body, an emotional body, and even a spiritual body. So, you can't treat one part without treating the whole in order to fully move something up and out. There is a physical component as much as there's a mental, emotional, and even spiritual energetic component. You can't just address one aspect and not the other. You have to support the physical body, especially when the trauma, negative energy, or emotional toxicity has been in there for a while. It's about treating the whole person, which is why SOUL INTELLIGENCE® aligns with functional and integrative medicine.

This experience taught me a profound lesson: our bodies are barometers of our emotional and spiritual states. When we ignore or suppress our emotions, they will eventually manifest as physical symptoms. True wellness can only be achieved when we honor and harmonize all of our parts. Healing requires us to address the whole person—mind, body, and spirit. This holistic approach opens the door to a more enriched, balanced, and fulfilling life and is at the heart of the SOUL INTELLIGENCE® method, which aligns closely with the principles of functional and integrative medicine.

Bioenergetics and Its Role in Healing

Bioenergetics, in its simplest form, is the study of how our mental and emotional states affect our physical health. It recognizes that we are holistic beings and that you cannot separate the mind from the body or the spirit from the emotions. When we experience stress or emotional turmoil, it impacts our physical health, just as physical illness can affect our mental and emotional well-being. When work says, "Leave your personal crap at the door at home and don't bring it in here to work," you can't separate yourself. You can't separate your mental, emotional, and physical body; it's all one you. Functional integrative medicine, which treats the entire body as a whole system—not just certain parts, functions, or merely symptoms—aligns perfectly with the principles of SOUL INTELLIGENCE®.

This approach goes beyond treating symptoms to addressing the root causes of disease, often incorporating innovative energetic therapies. For instance, light therapy, sound healing, and the SOUL INTELLIGENCE® method all work by shifting the body's energy to restore balance and promote healing.

Bioenergetics is the study of the connection between physical and emotional health. It's all about balance between mind, body, and spirit—our mental, emotional, physical, spiritual, and overall energetic health. When we are physically healthy, we are able to cope with stress and emotions. When we are emotionally healthy, our bodies are able to better stay in alignment and function properly. Just like we visit a chiropractor when we need a biomechanical adjustment. We need to energetically align our bodies with a bioenergetic massage.

Bioenergetics has been around since the 1950s, with pioneers like Alexander Lowen exploring the connection between physical ailments and unconscious psychic conflict. Lowen was a student of Wilhelm Reich, a psychoanalyst who believed and proved that physical ailments were rooted in unconscious psychic conflict. Lowen expanded on this idea to include the notion that chronic stress could also lead to mental health issues.

Today, the importance of mental and emotional health is more recognized than ever, particularly in the wake of the COVID-19 pandemic, which highlighted the profound impact of mental health on our overall well-being. We also experienced phenomena like "quietly quitting" and what was hailed as "the great resignation." People are leaving their jobs because they are no longer able to go on doing something they don't believe in, or something that doesn't hold meaning for them. The dissonance in frequency is too great.

In an era dominated by social media and constant connectivity, mental health challenges like anxiety and depression have reached epidemic proportions. The pressure to curate a perfect life online, coupled with the relentless pace of modern life, has led to a crisis in well-being. Mental health wasn't talked about openly when I was growing up. Most of the mental institutions, hospitals, and sanitariums are now closed, and we moved our mentally ill into our prison system. Many public figures are now speaking out about the need to focus on their mental and emotional well-being. Olympian Simone Biles spoke out about how she needed to leave the sport of gymnastics to work on her own mental health issues. Actress Kristen Bell has spoken out publicly about her battle with depression and anxiety and is an advocate for mental health. In fact, in 2020, depression was the number one cause of disability worldwide.

There is a strong connection between the mind, the spirit, and the body. To enjoy a wellness lifestyle, a person has to practice behaviors that will lead to positive outcomes in all dimensions of wellness. These dimensions are interrelated; one dimension frequently affects the others. For example, a person who is emotionally "down" often has no desire to exercise, study, or socialize with friends and may be more susceptible to illness and disease. We know that most diseases are a result of toxicity, stress, nutrient deficiency, and finally genes. Only 1% of all diseases are genetic, which means most of what goes wrong in the physical body can be addressed by changing our lifestyle and supporting our mental, emotional, and spiritual well-being.

Spiritual health provides a unifying power that integrates the other dimensions of wellness. Basic characteristics of spiritual people include a sense of meaning and direction in life, a relationship to a Higher Being, freedom, prayer, meditation, faith, love, closeness to others, peace, joy, fulfillment, and altruism. Studies have shown that spirituality strengthens the immune system, is good for mental health, prevents age-related memory loss, decreases the incidence of depression, leads to fewer episodes of chronic inflammation, and decreases the risk of death and suicide. A sense of joy, awe, and wonder has also been shown to lower inflammation and increase life expectancy. Imagine that!

The Holistic Approach to Wellness

Wellness is about balance—between mind, body, and spirit. To truly live a healthy life, we must nurture all dimensions of our being. When one aspect of our wellness is out of balance, it affects all the others. For example, when we are emotionally distressed, we may lose the motivation to exercise, socialize, or take care of ourselves, which can lead to physical illness.

As the limitations of conventional medicine become more apparent, there is a global shift towards holistic wellness, with millions of people seeking alternatives that address the mind, body, and spirit. This trend reflects a broader cultural movement towards personal empowerment, environmental sustainability, and a return to ancient wisdom. Spiritual health, in particular, plays a crucial role in integrating and unifying the other dimensions of wellness. A strong spiritual foundation provides a sense of meaning and direction in life, fostering qualities like love, peace, joy, and fulfillment. Research has shown that spirituality can strengthen the immune system, improve mental health, and even extend life expectancy.

By embracing a holistic approach to wellness, one that considers the physical, mental, emotional, and spiritual aspects of our being, we can achieve true health and harmony. The SOUL INTELLIGENCE® method provides the tools to access and align these different dimensions, helping us to live our best lives and fulfill our highest potential.

Journal Prompts

Think of the last time you didn't feel well, what do you think might have been the energetic root cause?

What unresolved trauma are you still carrying around that continues to affect you today

CHAPTER 3

CONSCIOUSNESS AND AWARENESS

The Levels of Consciousness

Consciousness is awareness, and there are different levels of consciousness that we can access and choose to operate from in our lives. It is the lens through which we experience life, and it operates on multiple levels, each offering a unique perspective on our journey. These levels of consciousness are not just abstract concepts; they are the stepping stones of personal growth and healing that guide us toward our highest potential.

The first level is where we get caught up in the drama of life—believing that life is happening to us. At this level, we are reactive and entangled in the emotional storylines of our experiences. It's easy to feel like a victim here, overwhelmed by the details, the cast of characters in the play, the hurt, and the perceived injustice.

As we evolve, we move to the second level, where we begin to see life happening for us. This shift in perspective allows us to find meaning in our challenges, to see them as opportunities for growth rather than mere obstacles. Most of us want to get to this level very quickly. As we begin to unpack the story, we see potential solutions. We can also begin to find a reason for the situation and perhaps incorporate some learning as earth is truly life university. Here, personal growth begins as we start to unpack the story and seek solutions to our pain.

The third level takes us even further. Before acting on our feelings, we step back and ask, "Who do I need to be in this situation?" This is where consciousness expands, and we begin to operate from a place of understanding that life is happening through us. We recognize the role we need to play to achieve the best outcome for all involved, and we start to embody the growth and wisdom that each situation offers. You feel the divinity of the situation and often see what growth is available for you.

This is the state of awareness that I embody when I'm doing SOUL INTELLIGENCE® work. The deepest level of consciousness is where we experience oneness with all. Where life is happening as we are. In this state, we transcend the self and slip into a connected awareness where we feel the butterfly effect—how our actions ripple out to affect the whole. This level is often accessed in moments of deep connection in spiritual practice or within nature, where the vastness of the ocean or the quiet majesty of a mountaintop allows us to feel part of something much greater than ourselves. It's an expansive state, a glimpse of the collective consciousness, but one we don't hold for long as we return to the duality of earthly life. It's a fleeting state but one that offers profound insight into the interconnectedness of all things.

Overcoming Self-Limiting Beliefs and Trauma

Our journey through consciousness is often hindered by self-limiting beliefs and unresolved trauma. These mental and emotional blocks are deeply entrenched in our subconscious, shaping how we perceive ourselves and the world around us. Imagine breaking free from the chains of past traumas that have silently dictated your life choices, relationships, and self-worth. What if the very thoughts that have held you back for years could be shifted in a way that changes your entire outlook on life?

My own experience with these limiting beliefs is a testament to their power. Growing up with an alcoholic and destructive father, I learned early on that overachieving was the only way to receive positive attention. I experienced a lot of fear and trauma growing up, as he was a very challenging personality to be around. It wasn't enough to meet expectations, but you needed to exceed them. To this day, I remember no matter what I did, I couldn't escape being compared to Jessica Marrin.

Jessica was actually a friend of mine, and her dad liked to brag about her at the local bar. My dad was always sure to tell me how Jessica earned straight "A's," so why couldn't I do that? This environment ingrained in me a deep-seated belief that I was not good enough, a belief that drove me to become an overachiever, constantly striving to prove my worth. It wasn't until I began to rewire my childhood programming through SOUL INTELLIGENCE® that I could truly embrace the reality that I am enough—more than enough. I realized that my worth is inherent, not something to be earned through relentless achievement. The trauma of my childhood also cultivated a predisposition to take care of others, a trait that would later define my professional life. From a young age, I felt responsible for my younger brother and often acted as the caretaker in our dysfunctional family, as my father was violent and drunk a lot. My mom had to work two jobs at one point to keep us afloat. I was always more concerned about others' well-being than I was about my own. I believe this predisposition of taking care of other people was where my healing nature originated.

Trauma is a complex, multi-layered experience that shapes our identities and influences every aspect of our lives. SOUL INTELLIGENCE® provides a profound exploration of how trauma lodges itself in our minds, bodies, and spirits, often dictating our thoughts, emotions, and behaviors without us even realizing it. The insights SOUL INTELLIGENCE® provides go way beyond surface-level understanding, giving clients the opportunity to get to the core of their pain, where true healing can begin. Our trauma often offers a pathway to liberation. I also believe that sports saved my life. It gave me something to excel at to earn some of the praise I sought so desperately. It also kept me out of the house after school and bought me more time away from the abuse of my father. I always enjoyed being a part of a team sport. I played field hockey, softball, skied, and swam, so I was always an athlete and liked being a part of a group. I enjoyed relationships and people very much and the synergy of being a part of a team working towards a common goal.

When I got out of school, what really resonated with me was when I started working for a college where I would go out and motivate high school students at 7am on goal setting, professional image, etc. How to select a career based on your personality type. I did Myers-Briggs stuff with them and did other things that fascinated me. At the end of the presentation, they would get a comment card, and they could tell me what they thought of the presentation. If they were interested in what the college offered, then I would bring those comment cards back, so that's really how I got into my career of being a turnaround specialist. I actually did that for colleges and universities and then for outside companies that would support colleges and universities.

It was fun. I was actually the first high school presenter in the company to get invited to the high achievers' conference that was typically reserved just for the salespeople who enrolled students in school. I got invited, and I did a whole presentation where I made the CEO and the entire executive team stand up, and I moved them around the room, and I put them in communication styles. It was a presentation I would do for high school students, and everybody had a great time. I remember that was where I was elevated within the company.

My background was really in higher education, and then it expanded into online learning, personal development, and leadership coaching. I was always helping people, and I thought the promise of higher education is so that you get to become what you really, truly want to become. It was really close to where I thought I was going to land, but I had no idea that this is what the Divine had in store for me at that time.

The archetype of the healer extended into my career, where I found fulfillment in helping others achieve their goals. But it also meant that I carried the burden of others' well-being, often at the expense of my own.

Overcoming these self-limiting beliefs is a challenge because they are so deeply rooted in our subconscious. We think the same thoughts over and over, and they form beliefs that cloud our perception of reality, like dirty sunglasses. For instance, if you believe you can't win, you'll subconsciously create situations that confirm that belief, leading to self-defeating outcomes. You may even put yourself in impossible situations just to prove yourself correct. If you believe you're not good enough, you'll constantly push yourself to do more, never feeling satisfied, even when you succeed.

The SOUL INTELLIGENCE® method allows us to dive deep into the roots of these self-limiting beliefs and allows us to transmute the energy and transform these invisible barriers that have held you back from achieving all that you desire. The impact is profound: clients often find themselves achieving goals they once thought were impossible, embracing opportunities they previously shied away from, and living lives that reflect their true potential. It's a life-altering shift from "I can't to I can and I will and I have already achieved it."

In the SOUL INTELLIGENCE® method, I use a chart with 25 common self-limiting beliefs. While there are certainly more, these are the most prevalent, and most other beliefs can be traced back to these core issues. I still sometimes struggle with old beliefs. Writer's block, for example, is often a manifestation of feeling not good enough. As I was writing this section, I felt bloated—a physical manifestation of self-doubt, self-sabotage, and not feeling good enough.

Recognizing this, I did a SOUL INTELLIGENCE® session on myself, followed by making my favorite shake to support digestion and some enzymes, addressing both the energetic and physical aspects of the issue. I assisted myself energetically and physically.

It's not one or the other; it's both. I also needed to look at the emotions around it, the feelings around the bloating. Emotion comes from the Latin word ex-movere, which means to move out. I moved the trapped emotion and self-destructive energy up and out of the body. It's important to remember that if your heart and mind are not in coherence, your subconscious mind will take over and manifest how you feel. Repetitive thoughts and self-limiting beliefs can wreck havoc if left unchecked. The SOUL INTELLIGENCE® method provides a way to identify and release these blocks, allowing you to move forward with clarity and purpose.

The Ever-Evolving Dance of the Divine Masculine & Feminine

As I reflect on my own spiritual journey, I believe my distorted views of the masculine and feminine dynamic really started in childhood. Growing up in the 1970s-1980s, the masculine was revered and the feminine was considered weak. When you consider that in the 1970s, we had women's rights begin to expand to include Title IX funding for sports, Roe v. Wade protecting a woman's right over her own body in pregnancy, housing discrimination on the basis of sex, and the Equal Credit Opportunity Act doing away with the practice of banks requiring a single, widowed, or divorced woman to bring a man along to cosign a credit application, amongst other decisions like no longer forcing pregnant women to take maternity leave, discrimination against women based on pregnancy, childbirth, or related medical issues, and the development of nonsexist teaching materials, or excluding women from juries.

I remember always feeling less than a man. While I was home for the summer from college, I had a job tending bar at a local restaurant. I felt the need to prove myself to the guy bartenders that I could do what they did just as well or better. So, I would struggle to carry two cases of beer up two flights of stairs–just to prove I could do it! I remember being so proud of the fact that I could actually do it, albeit winded. Seems silly now when I think back on it.

In my professional years, I also experienced the reality of "hitting the glass ceiling," knowing male colleagues were making more than I was doing the same job, or worse, I was doing a better job and still paid less. I felt devalued, so I believed that in order to make as much as they were, I had to do more. I had to be better; in fact, I had to be exceptional.

I also remember going through sexual harassment training in the early nineties and making a joke about it. I remember I put a welcome mat outside my office door, as sexual harassment was deemed inappropriate if it was "unwelcome." What is insane is that I had experienced sexual harassment throughout my career but thought that it was just part of the territory of working with men. What a warped view of masculinity I had at the time.

Criticism when you are a high achiever can really make you feel inadequate, scared, and like your job is always in jeopardy. It would bring up all kinds of irrational childhood fears of not feeling safe. And I sometimes still experience that fight, flight, freeze, or fawn trauma response to this day if I haven't met an expectation in some way. As my wise mom shared with me, focus on progress, not perfection.

One of the other obstacles that became a common theme and pattern for me was I had the uncanny knack (for marrying) and working for narcissistic men. Energy patterns, archetypes, and roles repeat personally and professionally —your life is your life, so in spite of our minds and egos wanting to think we can separate or compartmentalize, we can't. Energy is everything and everywhere.

Right before I got fired for the first time in my career in March 2020, I distinctly remember being in an executive meeting trying to explain to the CEO how the way we were rehiring people back into the facility after a reduction in force was causing a lot of unrest amongst the employees who stayed on with us. The reason there was such a backlash is that we were hiring their former colleagues back at a higher rate of pay. Our existing employees were really upset and it was negatively affecting performance.

The CEO just didn't want to hear it. He completely lost it on me and screamed at me like I was a child, reminding me what my job was and that it was my responsibility to handle the situation. He was cruel, condescending, wouldn't let me explain, he didn't care; those people should be grateful they have jobs; their colleagues were let go. It was really humiliating, as it was done in front of the entire executive team and my direct reports. My immediate supervisor was the COO, and he shared with me later that my best move was to just be quiet and take it. He gave me the analogy that the more I tried to explain, the more I appeared to be like the poor ice fisherman who had fallen into the cold lake, flailing in the ice, trying desperately to get out of the water and not drown.

According to the Mayo Clinic, the definition of narcissist personality disorder is as follows: it is a mental health condition in which people have an unreasonably high sense of their own importance. They need and seek too much attention and want people to admire them. People with this disorder tend to lack the ability to understand or care about the feelings of others, lacking empathy. They may also have a preoccupation with power, success, and/or beauty. They also exhibit manipulative behavior, including gaslighting, to get others to doubt their own perception of reality and question their sanity.

Needless to say, narcissists don't process emotions like empaths do—if at all. So, what is the definition of an empath? An empath feels. An empath is someone who is highly attuned to the emotions of others and may experience them on a deeper level than most people. The term comes from the word "empathy," which is the ability to understand someone from their perspective. Empaths are said to be able to detect subtle emotional shifts and unspoken feelings and may even "absorb" or "take on" the emotions of others. This can lead to deep relationships, but it can also make empaths susceptible to stress and emotional fatigue.

A story as old as time—the narcissist and the empath. They go together like peanut butter and jelly. Why? The narcissist is able to keep the empath in a cycle of emotional or physical abuse and continue to demoralize the empath and use them as the scapegoat for their own dysfunctional feelings. Empaths tend to internalize feelings and accept blame. The empath's strong sense of obligation towards others and fear of being perceived as "bad" causes them to work harder at being understanding and compassionate, resulting in an internal buildup of rage and occasional explosive behavior.

If the narcissist and empath form a personal intimate relationship, at first, sparks fly—it's all passion and intensity. But as the relationship unfolds, things get messy. The narcissist's need for control clashes with the empath's desire for authenticity, leading to power struggles and emotional rollercoasters. As you can imagine, the cycle between an empath and a narcissist can be difficult to break.

The men I worked for during most of my corporate career were narcissistic men—on the positive side—excellent in business but a disaster otherwise. All of these male figures in control of my life were reflections of the unresolved issues that I had with my own father. It's no wonder I suppressed my emotions for a very long time. I was told over and over again that you're not allowed to be emotional at work and never were you allowed to cry. Just like Tom Hanks' character says in the movie about a female baseball team, "A League of Their Own," "there's no crying in baseball."

For years, I was stuffing all of my emotions down inside, consistently invalidating myself as a woman, trying to be a more masculine figure. I tried to be "one of the guys." I actually made fun of anyone who cried at work. Looking back now, even my language choices emulated male behavior because that is what was accepted and encouraged.

I pushed down everything within me that was divinely feminine to such a degree and for so long that I experienced a cervical cancer scare. I went into my OB/GYN's office for an annual routine checkup. I got a call to come back into the office as the doctor had found that I had an abnormal pap smear and they wanted to do an in-office biopsy called a colposcopy, where they cut your cervix and send samples into the lab for further evaluation. The doctor's office called me back again and said that there were too many samples with abnormal cells on my cervix, so they recommended I have a conical biopsy which requires you to go under anesthesia in the hospital. I went through all the fear, doubt, and anxiety that anyone does who hears the "it could be cancer" diagnosis. Then I went into inspired action where I sought to discover the answer as to what was the root cause of my distorted cervical cells.

The cervix truly represents the seat of the Divine feminine within the body. The cervix is housed in the uterus, where our creativity as well as our sexuality reside. My cells were screaming at me! The energy had pooled long enough in the body to cause a disease, a dysfunction, a distortion in my frequency. It certainly got my attention.

From the time I got the phone call from the doctor's office to the time I had the conical biopsy scheduled at the hospital, I looked up every healer I could. I called everyone that I knew that worked with energy medicine of any kind. I did some of my own SOUL INTELLIGENCE® work on myself, but I also went to other healers, as sometimes you need to just receive. One healer finally gave me a clue as to what had caused the cells to mutate—she told me that I needed to embrace my Divine feminine. At the time, I was like, What the heck does that even mean? Do I need to dress more feminine? Wear my hair long? Dance? Be softer? Cry openly?

After meditation and reflection, I realized what it really meant was that I had been seeing the feminine aspects of myself as weak, not valuable, a liability. I needed to embrace my Divine feminine by embodying strong women like Joan of Arc, Mary Magdalene, and goddesses like Kali Ma and Sekmet. These inspiring women helped to lead people and were compassionate, smart, and strong.

I also worked with several different healers who did energetic acupuncture and provided esoteric messages. One body worker, healer, and friend told me that the cancer would be gone by the time I got to the hospital. I held onto that image and intention. For weeks, I saw my cervix healthy, I saw myself happy, and I was fully embracing the totality of who I am—Divine Feminine and Divine Masculine—whole within myself. It was finally time to go to the hospital and have the biopsy. I had the procedure done, and I remember the doctor coming back to me, looking kind of stunned, and her saying that it's gone. We must've cut it all out. I remember laughing and crying tears of joy at the same time. Maybe we did cut out some of the bad cells, but all of the mental, emotional, and spiritual work also contributed to the energy shift within my physical body.

I think like everything else in the world of duality we experience here on earth, to heal completely requires energy to shift within all of your bodies—mental, emotional, spiritual, and physical. You really have to go deeper into the mental and emotional component, and that's why SOUL INTELLIGENCE® aligns with functional medicine so well, since they treat the root cause of what's causing the dysfunction in the body.

Embracing my divine feminine energy has impacted my life and work since that scare in a lot of ways. The next pivotal moment was shortly after March 12, 2020—the day I was fired for the first time in my life. A couple of weeks into my work hiatus, I did a really impactful guided meditation. I remember coming out of that meditation with tears streaming down my face, knowing that I couldn't go back into a corporate environment again. I felt like I would suffocate if I did, and I had an inner knowing that it was time to bring SOUL INTELLIGENCE® forward.

The only way I could have received that guidance is by being in touch with my divine feminine. Looking back on it now, I know that I would have never left that toxic corporate environment— apparently, I needed to be catapulted out. That abrupt ending is probably one of the best things that could've happened to me, as devastating as it was at the time, but it allowed me time and space to get in touch with my Divine feminine. Further, it gave me the opportunity to hear the divine downloads that came with respect to developing the SOUL INTELLIGENCE® method and gave me courage to introduce it to the world.

Today there is a lot of discussion around the rise of the Divine Feminine and embracing the feminine aspects of ourselves. I'm all for that, but not at the expense of suppressing the divine masculine. Emasculation and suppression are not the answer, and it's my belief that extremes are never good. It forces separation and division, not inclusion and oneness. I believe happiness is truly about balance and wholeness within ourselves, thereby reflecting that in our outer world.

The Impact of Consciousness on Healing

Shifts in consciousness can lead to profound healing experiences. As we move through the levels of consciousness, we gain the ability to see beyond the immediate challenges and recognize the deeper lessons life is offering us. When we operate from a higher level of consciousness, we are no longer reacting to life; instead, we are co-creating our reality with the universe.

This shift in perspective is crucial for healing because it allows us to see our experiences, even the painful ones, as part of a larger journey of growth and transformation. By embracing this broader view, we can release the emotional and energetic blocks that have been holding us back, making room for healing on all levels— physical, mental, emotional, and spiritual.

Epigenetics, the study of how behavior and environment can cause changes that affect gene expression, provides a scientific basis for the power of spiritual healing. Research shows that our thoughts, emotions, and beliefs can turn certain genes on or off, influencing our physical health. This aligns with the spiritual belief that healing comes from within—by altering our inner state, we can activate the body's inherent ability to heal, bridging the spiritual with the scientific in a way that empowers holistic wellness.

The SOUL INTELLIGENCE® method is designed to facilitate these shifts in consciousness, helping you to access the deeper levels of awareness where true healing can occur. Whether you are overcoming self-limiting beliefs, healing from past trauma, or simply seeking to live more fully in alignment with your higher self, this method provides the tools and insights you need to navigate your journey with grace and clarity.

Journal Prompts

Has there been a self-limiting belief that has clouded your vision and limited your growth?

What level of consciousness do you tend to hang out in most ofthe time and is it serving you well and why or why not?

CHAPTER 4

UNPACKING A SOUL INTELLIGENCE® SESSION

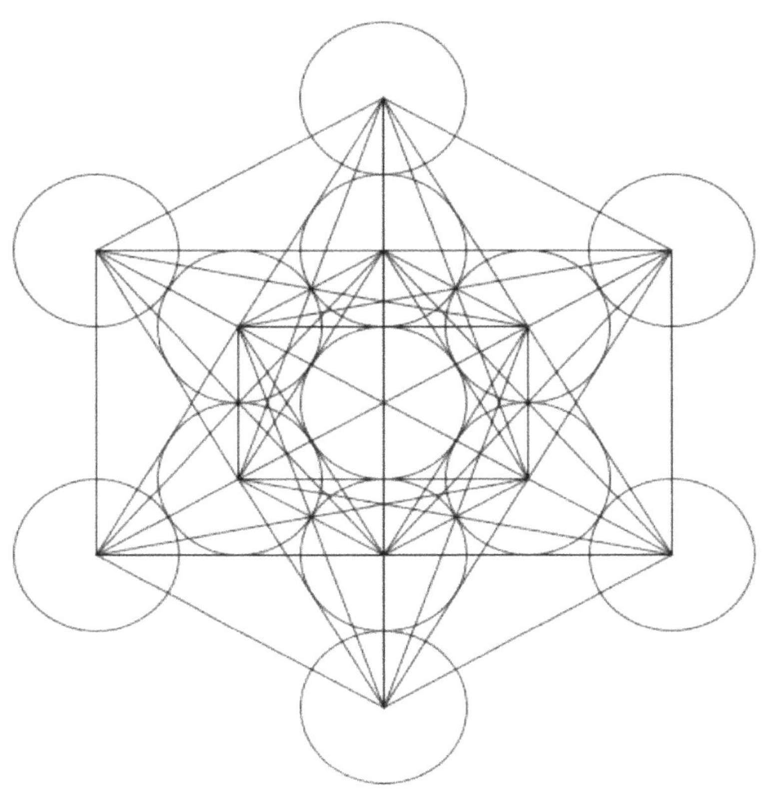

Preparing for a Session

The journey into a SOUL INTELLIGENCE® session, often referred to as an "energetic massage," is a deeply personal and transformative experience. One of the most beautiful aspects of this work is that there is no need for you to prepare in any particular way. In fact, the only requirement is your presence—whether in person or on Zoom. In a world where we are conditioned to believe that we must always be doing something, this session invites you to simply relax and receive.

Having an open mind and heart is ideal and can enhance the experience, but beyond that, the session unfolds naturally. We begin with a conversation—not a formal intake, but an open discussion about what's happening in your life. What's working? More importantly, what isn't? Whether it's a challenge at home, at work, or within a relationship, we explore it together. Life is all about relationships, so picking one of those that's not going according to plan is relatively easy. Often, the direction of the session is guided by the Divine, focusing on what will serve your best and highest good.

Once the theme of the session is established, I guide you into a more relaxed state to facilitate the healing process. You'll be invited to close your eyes, minimizing distractions, and focus on your breath. Through visualization, we connect with the Divine spark within ourselves, our connection to Source, grounding ourselves in the earth's energy while remaining open to the higher realms.

The Process of a Soul Intelligence® Session

Creating a sacred space is the first step in a SOUL INTELLIGENCE® session. I say a prayer on our behalf, setting the intention that anything we release during the session is transmuted by the Divine into love and light. This ensures that the energy we shift doesn't get trapped in our environment or affect those around us, including our pets, who are often sensitive to energy.

With the sacred space established, I begin by measuring your level of conscious awareness. Emotions correlate with different levels of consciousness and vibrations. Emotions like love, joy, and peace vibrate at the highest frequencies, while fear, guilt, and perfectionism resonate at the lower end of the scale. I use this measurement as a baseline, and we often see significant shifts as the session progresses.

The core of the session involves using a book of charts and a pendulum to identify the energies, emotions, and beliefs that need to be addressed. The pendulum, guided by divine information, helps pinpoint what needs to shift for your highest good. As we clear negative energies, toxic emotions, and self-limiting beliefs, we create space for supportive energies to flow in.

Once the clearing is complete, I check if there are any specific energies or supports that you need moving forward. This might include spiritual guidance, nutritional supplements, or other forms of healing and support that can help maintain the balance we've created.

We close the session with a prayer of gratitude, acknowledging the shifts that have taken place. Everyone reacts differently to an energetic massage. Some people may feel energized, but most people feel relaxed or even a little tired immediately after the session. However, the true magic often unfolds in the days and weeks that follow. That's when greater awareness and learning often occur. Once you have had a chance to let the new energy settle and review what's actually shifted, noticing what behaviors no longer trigger you and what pattern has been disrupted—therein lies the power of SOUL INTELLIGENCE®.

This new level of consciousness can be a slight change in awareness when you realize you are fascinated that you are no longer bothered by someone or something, or recognizing you don't feel the same way when experiencing a similar situation. Sometimes you might catch yourself in the moment and decide to take a different approach than you previously would have because you recognize that's an old way of being, doing, and operating in the world that no longer serves you. It's often a series of things that happen in terms of the awareness or consciousness levels changing that leads to behavior change. Ideally, you become a better leader, friend, spouse, or human who makes more positive changes as you go about your day. In general, I encourage you to pay attention to the changes in your life, noticing what no longer triggers you and how your relationships evolve.

The Energetic Massage: What to Expect

A SOUL INTELLIGENCE® session is often likened to an energetic massage—a process that releases the tension and blockages held within your energy field. Just as a physical massage works out knots in your muscles, an energetic massage releases trapped emotions and energies from your cells. This work is subtle yet profound, with each person reacting differently during and after the session.

To begin, it's important to create a conducive environment, free from distractions. Silence your phone, close other programs on your computer, and ease back into your chair. Close your eyes, allow your jaw to become slightly slack, and allow your shoulders to drop away from your ears. One of the greatest gifts you can give yourself during this time is your undivided attention, embracing the silence as you tune into your body and spirit.

As we start the session, I guide you through a visualization to connect with your divine spark, expanding your light until it fills the room. Imagine a brilliant white light move beyond your chest, expanding to your head and torso, then continuing to expand into the room—really take up space. Let this be a gentle reminder of what a truly powerful spiritual being you truly are, having a human experience. Draw all of that expansive energy back into your physical body and see it form a column of light going up your spine through your hips, your stomach, your chest, your throat, between your eyebrows, to the top of your head. See that column continue beyond the crown of your head to about 12 inches above. This is known as the Twelfth Gate; that's where your higher self resides, that best version of you. The one that is so proud of how far you have come and is so excited about the journey that still lies ahead. See yourself making a heart-to-heart connection with that being.

See your light extend out into the sky, passing the sun, the clouds, stretching further into the atmosphere, passing the stars, the moon, then moving into the galaxy, passing the planets and comets. Further expanding into the center of the universe to merge with the most brilliant white light you've ever seen. Feel as though you are a part of that light; join that light; allow it to envelope you.

Pull that light all the way back down through the top of your head, down through your physical body, exiting your root, going down your legs to your feet, connecting you with the crystalline grid of the earth. Connect with Mother Nature, Gaia, the ground, the dirt, rocks, sand, crystals, underground oceans, rivers, and middle earth, all the way to the lava core, and drop a grounding cord, anchoring yourself. Then feel all that supportive earth energy coming back up through the layers of the earth, through your feet, your legs, into the center of your physical body, all the way to your heart. Our heart chakra is where the heavenly and earthly energies commingle. Feeling fully supported both above and below, we then say a prayer of protection.

This process reminds you of your true power as a spiritual being having a human experience. We draw this expansive energy back into your body, aligning it with your higher self—the best version of you, who is ever proud of your journey and excited for what lies ahead.

From there, we connect with the energy of the earth, grounding ourselves in its supportive embrace. We then say a prayer of protection, surrounding ourselves in a bubble of white light and inviting our spiritual support teams to join us. With everything aligned energetically, we begin the process of identifying and shifting the energies and emotions that no longer serve you.

Using a book of more than 30 charts and a pendulum, we dive into the detailed aspects of your energy field. The pendulum helps to discern what needs to move, shift, or be clear, as the charts are very detailed. The first chart serves as a table of contents, and that's where you always begin from, as this informs me as to what chart to use. The charts cover a wide range of issues, from primary relationship challenges to traumatic imprints, hereditary issues, physical illness, and beyond. We might start with one of those issues, and then I might go a little deeper and investigate more details like when this occurred, who was involved, who caused harm, etc. We explore the specific energies or emotions at play, whether they are negative thoughts, destructive repetitive patterns, toxic masculine and feminine energies, non-supportive archetypes, or self-limiting beliefs. The goal is to clear these blocks, making space for healing and growth.

Once we have made space, I check to see what supportive energies need to be brought back in. It could be aligning your energy with wisdom, light, and understanding or something more specific where you might choose to also take it in the physical, like a Bach flower remedy or a nutritional supplement. As we conclude the session, I invite you to ask any questions or seek guidance on specific decisions you may be facing. The final step is a prayer of thanks, sealing the work we've done with gratitude.

What to Expect After a Session

Immediately after a SOUL INTELLIGENCE® session, you might feel a range of sensations —some people feel deeply relaxed, while others are energized. But the real transformation often reveals itself over time. As the days and weeks pass, you may notice subtle yet significant shifts in your life. Perhaps something that once triggered you no longer has the same power, or a relationship that felt strained now feels more harmonious.

The session recording is available for you to revisit, and you are encouraged to reflect on the changes you observe. This ongoing awareness is where the true power of the session lies, as you continue to integrate the shifts into your daily life. A SOUL INTELLIGENCE® session is more than just a healing experience —it's a profound reconnection with your true self, a step toward living fully in alignment with your soul's purpose. By clearing away the blocks that have been holding you back, you open the door to a more expansive, fulfilling life.

Journal Prompts

How often are you willing to give yourself the natural gifts of silence, solitude, stillness, and simplicity by just being?

If you were able to schedule a SOUL INTELLIGENCE®session, what would you want to shift energetically and why?

CHAPTER 5

INTEGRATION AND DAILY PRACTICE

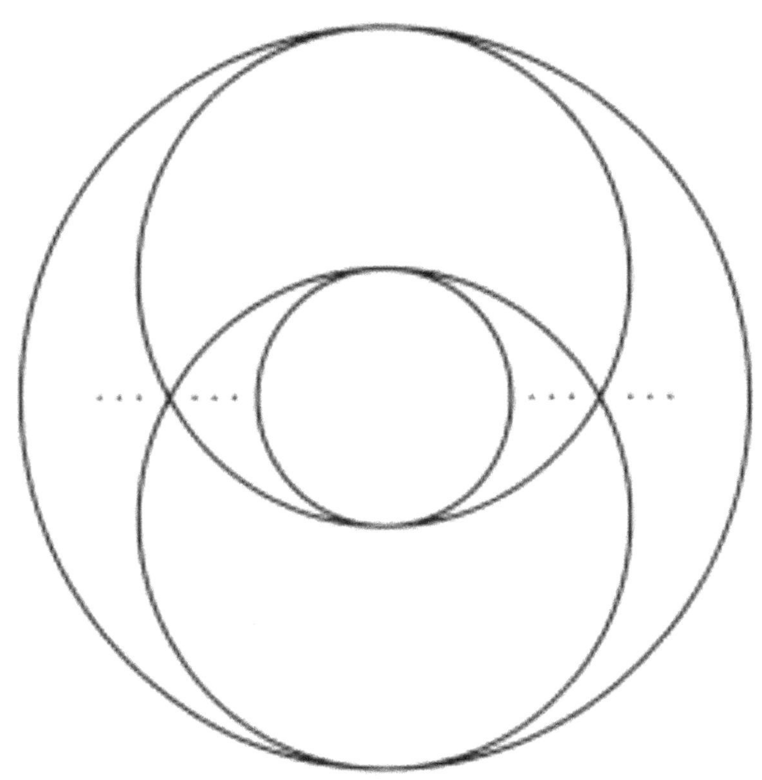

Integrating Soul Intelligence® into Your Life

There are a lot of ways to allow your life to be led by your SOUL INTELLIGENCE® and tools you can use to integrate it into your way of being in the world. Incorporating the principles of SOUL INTELLIGENCE® into your daily life is not about adding complexity but rather about simplifying your approach to decision-making and aligning with your inner wisdom.

Imagine waking up each day with a clear sense of purpose, vitality, and inner peace. SOUL INTELLIGENCE® is a blueprint for lasting transformation in your everyday life. By integrating these practices into your routine, you're not just improving your well-being; you're cultivating a sustainable way of living that aligns with your deepest values and aspirations. Clients have reported significant improvement in their mental clarity, emotional balance, and overall happiness, making these practices a cornerstone for lifelong well-being.

One powerful technique I often use in my coaching practice is called "inside-out problem solving." This method helps you make decisions by tuning into how each potential outcome feels within your body, allowing your inner wisdom to guide you. Here's how it works: Stand in a fixed spot, close your eyes, and focus on a decision you need to make. Visualize several possible outcomes and consider each one carefully. If you choose a particular path, how does it feel in your body? How does it resonate within you? Project yourself one year into the future—how does that decision impact you? How does it influence others around you? Did it create the desired outcome you sought?

Then, take a physical step to the side and repeat the exercise for another potential outcome. If you have more than two choices, move to a third spot. By feeling into each option and visualizing the outcomes, you can connect with your SOUL INTELLIGENCE® and arrive at a decision that truly aligns with your highest good.

Remember, all the answers you seek are already within you. The key is to get quiet, eliminate distractions, and truly feel into each decision to know which direction resonates best. The Sacred Soul Script One of the most effective ways to integrate SOUL INTELLIGENCE® into your daily routine is through a practice I call the Sacred Soul Script. This journaling exercise helps you align your goals with your deeper desires, allowing your life to be led by your SOUL INTELLIGENCE®.

1. Begin by writing down a goal that you want to achieve in the present tense. It should be something that excites you and perhaps scares you a little. Choose a timeline that feels right—three months, six months, a year.

2. For example, "By 2025, I will be leading the SOUL INTELLIGENCE® *movement full-time, speaking across the country, sharing the power of this method, and teaching others how to incorporate this incredible biohack into their practices, businesses, and lives. "*

3. Next, write 3 to 5 beliefs that will support you in achieving this goal. These might be affirmations like, *"I am incredibly talented. I am an extraordinary healer. Everything always works out for me."*

4. Then, consider who you need to be to embody these beliefs. Write statements such as, *"I am the best version of myself. I am a confident, compassionate, aspirational, and conscious leader ready to lead the SOUL INTELLIGENCE® movement."*

5. Now, here's where the magic happens —write your goal as if you've already achieved it. Imagine it's a year from now: How do you feel? Who are you with? What did achieving that goal do for you, your family, your friends, and your colleagues? What trips did you take as a result of this success? Use as many feeling words as possible because your subconscious mind responds to emotions, feelings, and beliefs. To collapse the particle wave from the spark of your goal, you need to magnify it and draw it into reality using the power of emotion, feelings, and belief.

6. Next, list your top three core values and reflect on your goal. Is it in alignment with your values? Do you need to adjust your goal to better align with your mission, vision, values, and purpose?

7. For example, my personal values are to be authentic —true to myself; to inspire —others; and to enlighten —by helping others become lit up from the inside out. It's much easier to achieve a goal that aligns with your core values.

8. Finally, consider what you already have in place that reassures you that your goal is achievable. For example: *"I already have everything I need within me to achieve this*

goal. I have support from incredibly talented people who believe in me. I'm already teaching hundreds of thousands my SQ method."

9. The most important part of this practice is to read and visualize your Sacred Soul Script every morning and night. By doing so, you are using the 5% of your conscious mind to program the 95% of your subconscious mind. In essence, you are brainwashing yourself for good. More on that in the next chapter.

The Power of Daily Practices

Daily rituals such as meditation, visualization, and affirmations are essential tools for aligning with your SOUL INTELLIGENCE®. Setting aside time to embrace the quiet and silence is crucial. Whether through meditation, prayer, journaling, or reading inspiring messages, these practices create space for you to hear, see, and feel messages from the Divine.

Meditation is known for its calming effects, helping you reduce stress and improve focus. It doesn't need to be a big production, either. I often meditate before getting out of bed in the morning, so that I'm ready to take on the day before my feet touch the floor. Using guided meditation can also be helpful if you are a "Type A" personality who has trouble slowing down.

Below is a very simple daily intention setting meditation practice that you can follow:

1. **Find a quiet space:** Choose a place where you won't be disturbed and close your eyes.

2. **Focus on your breathing:** Take deep, slow breaths, inhaling through your nose, exhaling through your mouth. Try inhaling for four counts, holding for four counts, and releasing the air for eight counts. Repeat this at least three times to regulate your nervous system.

3. **Relax:** Allow your shoulders to drop away from your ears, allow your jaw to become slack, feel the chair rise up to meet you, so you feel fully supported.

4. **Be Present:** Put everything you were thinking about in a virtual box outside of your space, put everything you have yet to do today in that same box, so you can be fully present.

5. **Visualize:** Picture a brilliant white light in the center of your chest, growing with each breath, moving to encompass your body, then expand beyond your body, feeling into your true spiritual nature.

6. **Set Your Intention:** "Today I embody joy in every action I take. I flow through the day with grace and ease. Everything is always working out for me." Use feeling words to engage your subconscious.

7. **Stay Present:** If your mind wanders, gently bring your focus back to your breathing and visualization.

8. **Give yourself this gift daily.**

Meditation, once considered purely a spiritual practice, is now backed by neuroscience. Studies have shown that regular meditation can increase gray matter in the brain, particularly in areas associated with memory, empathy, and stress regulation. This tangible evidence bridges the gap between ancient spiritual practices and modern scientific understanding, proving the mind's ability to heal itself is rooted in both tradition and biology.

One of my absolute favorite meditations is quick and is guaranteed to put a smile on your face in moments. Close your eyes and focus on your heart. Now I want you to picture in your mind's eye someone, something, or some place that you absolutely love. It could be your lover, your dog, or an exotic island—or you are on an exotic island with your lover and your dog—your choice. The important thing is to fall in love. Another powerful practice is immersing yourself in nature—listening to the waves, hearing birds sing, or feeling the wind on your face. These moments of connection with the natural world help to align your frequency with a higher vibration. In order to align with your SOUL INTELLIGENCE®, you have to reduce your distractions and be available to hear, see, and feel messages from the Divine.

One of the ways I shift my energy up when I'm down is by helping someone else. When I give, I naturally receive. Even if it is something small, like giving change to a kid outside the grocery store who is collecting for his team. Helping an elderly person carry their packages to the car. Another way to shift your vibration is to focus on gratitude. You don't necessarily need to write out a list; I typically only do that if I'm really down. What about incorporating small moments of gratitude when you are out and about? Imagine standing in line at the UPS store, feeling the frustration of a long wait. Instead of stewing in irritation, take a moment to envision three things you are grateful for. This simple shift in focus can transform a mundane and potentially negative experience into one that uplifts your mood and sets a positive tone for the rest of your day.

I once set an alarm mid-morning and a second one mid-afternoon to remind myself to be authentic, soul-led, and confident. It was a fun mindfulness reminder to keep myself centered, grounded, and in touch with my SOUL INTELLIGENCE® during the work day.

Often when I'm delayed at the airport, I will do an intentional chocolate mindfulness meditation. That's where you indulge your senses —purchase a favorite chocolate candy. Put it in your hand, lift it to your nose, smell the chocolate. Close your eyes, and put it in your mouth. Savor it. Allow it to dissolve slowly, notice the sensation as it melts on different parts of your tongue. See how long you can resist chewing and swallowing it, allowing the experience to last. How can you possibly be annoyed with a flight delay when you are practicing chocolate "blissipline" like this?

Physically, how are you caring for the house where your SOUL INTELLIGENCE® resides? Do you nourish your body with healthy foods, supplements, and exercise? What are you doing to help your physical body operate at its best level? Do you incorporate small changes in your day, like drinking enough water? Try starting your day with an eight- or sixteen-ounce glass of water. We should be drinking half our body weight in ounces of water daily. Perhaps when you drink the water you can say an affirmation and feel it as the water goes into your body? Something like, "I am happy, healthy, and wealthy in all ways."

How do you manage toxicity in your environment, from the food you eat to the products you use on your skin and in your home? Are you eating grass-fed proteins? Or are they full of hormones? Take a look at what you put on your skin—the largest organ in your body. What kinds of household cleaners do you use? Are you using chemicals to kill weeds in your yard or to keep it green? All of these things are ways to keep your energy high and your body healthy. We can't all go to Tibet or live in a Blue Zone, so we need to learn to manage our energy and the toxicity that we expose ourselves to on a physical, mental, emotional, spiritual, and energetic level.

Remember, while you can't always control what's happening around you, you can control how you respond. There's a lot that we can do and that we control. Most of us feel like we can't control what's going on around us. In essence, you can't always control what's going on around you, but you can control how you respond or react. Am I going to give energy to this or not? Your response determines your energy and, ultimately, your reality.

The other thing that I have become even more aware of and better at mastering during the day is feeling into my body and becoming more aware and acknowledging of my feelings throughout the day. They really are the fuel of life. They bring in and magnify what begins as a thought and sometimes can become a belief, and we need to feel.

That's a big part of being human—to feel, to express, and to experience. Instead of trying to figure it out, I think we need to shift and feel it out. There's a practice called somatic therapy that helps you feel your emotions. Getting quiet and asking yourself, What is this ache in my heart really telling me? How does it feel? What does it look like? What's underneath that?

I would never think about taking a timeout during the day and meditating for 20 minutes. In my general corporate career, that's so counterintuitive; why would I do it? I realize if that thought comes, there's a reason for it, and I need to pay attention to it. I have to honor my body. We all talk about self-care; what does that mean? Is it selfish? You have to take care of yourself before you can take care of anybody else; otherwise, you're depleted and drained. I think you have to decide, what does that mean for me? For some people, that means they need to take a nap for 20 minutes in the afternoon. For other people, that might mean going to a spa and having some mega-self-care done.

Tools for Maintaining High Vibration

Maintaining a high vibrational state is key to living in alignment with your SOUL INTELLIGENCE®. In addition to the daily practices mentioned earlier, consider incorporating tools that resonate with you. One of my favorite ways to help me align my frequency to a higher vibration is to listen to high-vibrational music. It helps me shift very quickly. Sometimes I listen to music from the 80s and 90s, as it brings up fun memories for me. Another way to immediately shift your vibration higher is to laugh. It's probably my favorite way to shift up into joy.

Mindful company is particularly important. Who are you spending your time with? Are they elevating your energy or draining it? Consider who you spend the most time with on a daily basis. Are these people who support your goals? Are they excited for you? Do they inspire you? Or are they energetic vampires sucking the ever-loving life out of you? Do they question your sanity? Do they drag you down? Do you feel drained after engaging with them?

Consider who is helping you align with who you truly are and who enables you to live by your SOUL INTELLIGENCE®. We are only as good as the people we surround ourselves with on a regular basis. I know from tennis that I always play better when I'm playing up a level. You rise to the occasion.

Choose to be around those who inspire you and help you live in alignment with your highest self. Pay attention to what you're feeding your mind. Are you consuming content that uplifts and inspires, or are you allowing negativity to infiltrate your thoughts? I'm selective about what I read, watch, and listen to because it all effects and goes into my subconscious mind and can affect my energy and well-being. I read positive news and fun fictional novels for pleasure. If I indulge in TV, I typically watch sports or an occasional movie. Be intentional about the media you consume, the conversations you engage in, and the environments you inhabit.

Finally, honor your body as the vessel for your SOUL INTELLIGENCE®. Nourish it with healthy foods, supplements, and practices that support its optimal functioning. By maintaining a high vibration physically, mentally, emotionally, and spiritually, you create the ideal conditions for your SOUL INTELLIGENCE® to guide you in every aspect of your life.

Journal Prompts

What is one daily practice you are willing to incorporate to tap into your SOUL INTELLIGENCE®?

Is your inner circle one that elevates you, supports you, and inspires you to be the best version of yourself?

If yes, journal about what you value about their love and support. If no, consider why you continue to keep them close? What negativeimprint or belief are they reinforcing.

CHAPTER 6

BRAINWASHING FOR GOOD

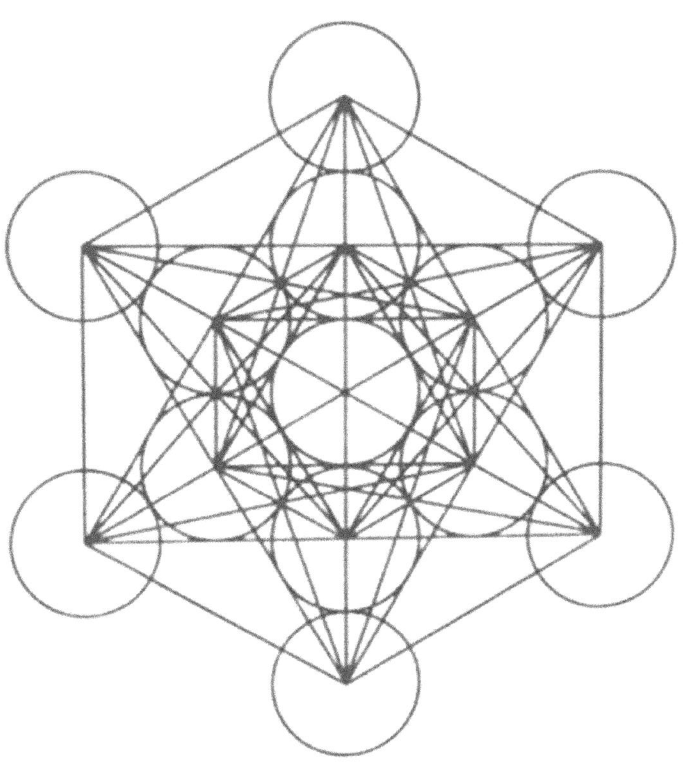

Understanding Brainwashing Techniques

The term "brainwashing" often evokes fear and mistrust, conjuring up images of manipulation, coercion, and control. Historically, it has been associated with harmful practices that strip individuals of their autonomy, influencing them to adopt beliefs and behaviors against their will. However, when we shift the narrative, brainwashing can be redefined and repurposed as a tool for positive self-transformation. When applied with the right intent, brainwashing becomes a powerful method for reprogramming the subconscious mind to help achieve your highest goals and aspirations. In essence, it's about brainwashing yourself for good— aligning your subconscious with your conscious desires to create meaningful and lasting changes in your life.

To understand this, it's essential to grasp the profound influence of the subconscious mind. Only about 5% of our daily mental activity is conscious, meaning that the remaining 95%—the subconscious— dominates our thoughts, beliefs, and behaviors. This vast and often untapped part of our mind holds the blueprint for our deepest fears, desires, habits, and beliefs. By effectively reprogramming the subconscious, we gain access to immense power, allowing us to break free from limiting beliefs and negative patterns and adopt new, empowering mental frameworks.

So, how does positive brainwashing work, and why is it effective? At its core, brainwashing leverages principles that have been used for centuries, sometimes in negative or coercive contexts, to influence the thoughts and behaviors of individuals. These techniques include isolation, repetition, control of messaging, and emotional reinforcement. But what if we could flip the script and use these same techniques to empower ourselves rather than to control others?

This chapter will explore the key steps and methods to engage in positive brainwashing, including various subconscious reprogramming techniques that can transform the way you think, feel, and act. We will also explore somatic therapy as an alternative and complementary method to further accelerate the reprogramming of the subconscious mind.

Isolation and Focus: Setting the Stage for Change

One of the foundational principles of brainwashing is isolation. In negative contexts, isolation is used to weaken an individual's connection to their familiar reality, making them more susceptible to external influence. But in the context of self-reprogramming, isolation serves a different purpose: it creates space for deep introspection and focus. This isolation doesn't mean cutting yourself off from the world; rather, it's about dedicating time and space where you can be fully present with yourself, away from distractions, influences, and obligations.

To begin the process of positive brainwashing, you should create a quiet, controlled environment where you can focus solely on the work of self-reprogramming. Set aside a designated space—perhaps a quiet room, a meditation corner, or even a retreat setting—where your primary goal is to realign your mind. During this time, your focus is on yourself and the messages you wish to implant into your subconscious. This space becomes sacred—a place where you're fully committed to your inner work. In this mental and physical isolation, you engage deeply with your thoughts and emotions, allowing your subconscious to become more receptive to change. The isolation acts as a buffer between you and the outside world, helping you disconnect from negative influences, distractions, and old patterns.

Control the Messaging: You Are the Creator of Your Reality

Once you've established your sacred space, the next step is to take control of the messaging that you feed into your subconscious. In traditional brainwashing, the manipulator controls the messaging, often bombarding the individual with specific information to alter their beliefs. In self-brainwashing, you become the creator and curator of the messages you wish to program into your mind. The subconscious mind is incredibly impressionable, and it absorbs information based on repetition and emotional resonance. Therefore, the messages you choose must be aligned with your goals, desires, and vision of the future. The messaging you use must be something that excites you, something you believe is possible to achieve. This messaging takes the form of affirmations, visualizations, and self-talk. But it's essential that these messages not only resonate with you on an intellectual level but also emotionally excite and inspire you.

To make these messages effective, they must feel achievable and authentic. You need to believe, deep down, that these messages are possible to realize. Most importantly, it must resonate with you so deeply that you can see it as already done. Your frequency has to match that which you believe you deserve. Therefore, you resonate with the feeling of joy that comes from having already achieved that goal. The more specific and personalized the messaging, the more potent it becomes.

Visualize your goals as already accomplished. This is where your vibration comes into play—your emotional and mental frequency must match the feeling of having already achieved your desired outcomes. For example, instead of simply telling yourself, "I will be successful," imagine yourself already living in that success. Picture the specifics: where are you, who are you with, what are you doing? Engage your senses—how does it feel, smell, and sound to be in this reality? Your subconscious doesn't understand the concept of time; it operates in the present moment. Therefore, your messaging must reflect the here and now, as if you're already living in the reality you desire.

Repetition with Feeling: The Key to Lasting Change

One of the most well-known brainwashing techniques is repetition. When an idea is repeated enough times, it begins to take root in the subconscious mind, eventually becoming a belief. But repetition alone is not enough to create meaningful change—it must be accompanied by emotion. Your subconscious is deeply influenced by feelings. Every thought you have that is infused with emotion becomes more powerful and impactful. The reason for this lies in how the brain processes emotions. When you experience strong feelings—whether positive or negative—your brain releases neurotransmitters that heighten the intensity and memorability of the associated thought. This is why emotionally charged memories, whether joyful or traumatic, tend to stick within our cells more than neutral experiences. To brainwash yourself for good, repetition must be combined with intense emotional engagement. The more emotion you infuse into your repetition, the more powerful it becomes.

Whether you are repeating affirmations, visualizing success, or engaging in self-talk, make sure that you are infusing these practices with positive emotions such as joy, love, gratitude, and excitement. For example, when repeating an affirmation like, "I am confident and successful," truly feel what it's like to be confident and successful. Imagine standing tall, speaking with authority, and being embraced by others for your leadership.

As you continue to practice, the emotional charge strengthens the new beliefs you're cultivating. These beliefs are then stored in your subconscious as truths, and over time, they begin to guide your actions, thoughts, and decisions, leading to real-world manifestations of your internal programming.

Somatic Therapy: Releasing Trauma and Rewiring the Body

While affirmations, visualization, and repetition are powerful tools for subconscious reprogramming, there is another dimension that often goes overlooked—the body. The subconscious mind is not only stored in your thoughts and emotions but also within your body. Physical sensations, tension, and trauma are often held in the body, shaping our beliefs and behaviors on a deeper, often unconscious, level. This is where somatic therapy comes into play. Somatic therapy is a body-centered approach that helps individuals release stored trauma, tension, and emotions from their bodies, allowing for deeper healing and subconscious reprogramming. By focusing on physical sensations and the mind-body connection, somatic therapy works to unlock and release the subconscious patterns that are held in the body.

How Somatic Therapy Works

Somatic therapy involves paying attention to the sensations in your body and working with them to release stored emotions and trauma. Often, traumatic or highly emotional experiences are stored as physical sensations—tightness in the chest, knots in the stomach, or tension in the shoulders. These physical manifestations of stress or trauma can block your ability to fully reprogram your subconscious mind, as they hold onto outdated beliefs and emotions.

Through gentle physical exercises, body awareness, breathwork, and mindful movements, somatic therapy helps to release these trapped energies. When you let go of these stored emotions, the subconscious becomes more malleable and open to new ideas and beliefs. By healing the body, you create a clear channel for reprogramming your subconscious mind.

Integrating Somatic Therapy with Positive Brainwashing

Somatic therapy can be easily integrated with the other subconscious reprogramming techniques discussed. Here's how:

1. Body Awareness During Visualization

While engaging in your daily visualizations or affirmations, take a moment to become aware of the sensations in your body. Do you feel any tension or resistance? Is there tightness in certain areas, such as your chest or shoulders? Pay attention to these sensations and allow yourself to breathe into them, releasing any tension as you focus on your positive outcomes.

2. Breathwork for Releasing Stored Emotions

Incorporate deep breathing exercises during your affirmations or self-hypnosis practice. Focus on deep belly breathing, inhaling fully and exhaling slowly. This helps release stored emotions and tension, calming the nervous system and creating a relaxed state where the subconscious is more open to new ideas.

3. Movement to Unblock Energy

Simple, mindful movements can help release blocked energy in the body. This might include gentle yoga, stretching, or shaking exercises. As you move, focus on the areas where you feel tension or resistance, visualizing that tension melting away with each movement. When the body is relaxed and free from tension, your affirmations and visualizations can penetrate deeper into the subconscious mind.

4. Releasing Trauma to Clear the Path for Positive Beliefs

Trauma held in the body can create mental blocks and resistance to change. By working with a somatic therapist or practicing somatic exercises on your own, you can begin to release these old traumas, clearing the way for new, empowering beliefs to take root. As you release these physical and emotional blockages, your subconscious becomes more receptive to positive brainwashing techniques.

Advanced Subconscious Reprogramming Techniques

In addition to the basic framework of isolation, messaging, repetition, and love, and the incorporation of somatic therapy, there are several advanced subconscious reprogramming techniques that can further enhance the effectiveness of your brainwashing practice. These methods tap into different aspects of the mind to deepen the process of transformation.

1. Hypnotherapy and Self-Hypnosis

Hypnotherapy is a powerful tool for accessing the subconscious mind in a direct and focused way. Under hypnosis, your conscious mind relaxes, allowing you to bypass the mental filters that often block new beliefs from taking root. During a hypnotherapy session, a trained therapist guides you into a deep state of relaxation and then introduces positive suggestions or affirmations into your subconscious.

Self-hypnosis works in a similar way, but you are in control of the process. You can learn to induce a state of relaxation through meditation or breathing exercises and then repeat your affirmations while in this state.

The relaxed mind is more receptive to new ideas, making self-hypnosis an excellent tool for reprogramming the subconscious.

2. Neuro-Linguistic Programming (NLP)

NLP is a psychological approach that involves reprogramming the way you process information. Through a series of techniques such as anchoring, reframing, and pattern interruption, NLP helps you change your internal dialogue and mental patterns. For example, if you have a limiting belief like, "I'm not good enough," NLP can help you identify where this belief originated and replace it with a more empowering one. One technique is called "anchoring," where you associate a specific physical action (such as pressing your thumb and forefinger together) with a positive emotional state. By doing this repeatedly, you train your mind to automatically access that positive state whenever you perform the anchor.

3. Binaural Beats and Brainwave Entrainment

Binaural beats and brainwave entrainment are audio tools that can alter your brainwave patterns, helping you access different states of consciousness. For reprogramming the subconscious, you want to access the theta brainwave state —a state of deep relaxation and heightened suggestibility. Binaural beats are sound frequencies that, when listened to with headphones, guide your brain into specific brainwave states. Listening to binaural beats while repeating affirmations or visualizing your goals can enhance the effectiveness of your practice by making your subconscious more open and receptive to new ideas.

Reprogramming Your Subconscious for Success:

A Step-by-Step Guide: Reprogramming your subconscious mind requires consistent practice, but the steps are simple and accessible to everyone. Below is a practical step-by-step guide to brainwashing yourself for positive outcomes using these techniques:

1. Create Your Script

Start by writing down your goal as if it has already been achieved. Use the present tense and incorporate as many sensory details as possible. How does it feel to have achieved this goal? Who are you with? What are you doing? Engage all your senses to make this vision as vivid as possible.

2. Affirm Your Beliefs

Identify the beliefs that will support your goal. Write down three to five affirmations that reinforce these beliefs. These should be statements that reflect the person you need to become to achieve your goal, such as, "I am confident, powerful, and successful."

3. Visualize Daily

Twice a day, take a few minutes to read your script and affirmations. As you do, close your eyes and visualize your goal as if it's already reality. The key here is to immerse yourself in the feelings associated with achieving your goal. Feel the joy, pride, and satisfaction as if it's happening right now.

4. Reinforce with Love

After each visualization, take a moment to reward yourself with love. This could be as simple as smiling, placing your hand over your heart, or saying something loving to yourself. Immerse yourself in the feeling of gratitude and joy for having already accomplished your goal. The act of rewarding yourself with love helps to anchor these new beliefs in your subconscious.

5. Consistency is Key

The effectiveness of this practice lies in consistency. Make it a non-negotiable part of your daily routine. Over time, these repeated thoughts and feelings will become ingrained in your subconscious, guiding your actions and decisions in alignment with your goals.

The Power of Repetition, Feeling, Love, and Body Awareness

- The triad of repetition, feeling, and love is crucial in reinforcing positive change within your subconscious mind. However, adding body awareness through somatic therapy can deepen this process even further. Each element plays a specific role in this reprogramming process:

- **Repetition**: The subconscious mind learns through repetition. The more you repeat a thought, the more it becomes a belief. You are changing your internal condition and creating new neural pathways. This is why daily practice is essential —it ensures that your new, positive beliefs take root and flourish.

- **Feeling**: Emotion is the language of the subconscious mind. When you infuse your repetitions with strong, positive emotions, you amplify their impact. It's not just about saying the words; it's about feeling them with your entire being. This emotional charge helps to bypass the conscious mind and directly influence the subconscious.

- **Love**: Love is the most powerful force in the universe, and when used in this context, it acts as a potent reinforcement. By rewarding yourself with love, you create a positive feedback loop that encourages your subconscious to accept and embrace the new beliefs you're instilling.

- **Body Awareness**: By releasing stored trauma and tension through somatic therapy, you clear physical and emotional blockages that may be preventing your subconscious from fully accepting new beliefs. This creates a holistic and integrated approach to reprogramming, allowing both mind and body to align in the process of transformation.

- By understanding and applying these principles, you can effectively brainwash yourself for good, aligning your subconscious mind with your highest intentions and creating the life you truly desire.

Journal Prompts

If you could successfully brainwash yourself for good, what is one area of your life you would like to focus on?

What are some ways that you could reward yourself with love?

Where in your body do you feel tension or resistance when you think about your goals? How can you release that tension throughmindful movement or breath?

CHAPTER 7

THE INTERSECTION WITHHEALTHCARE

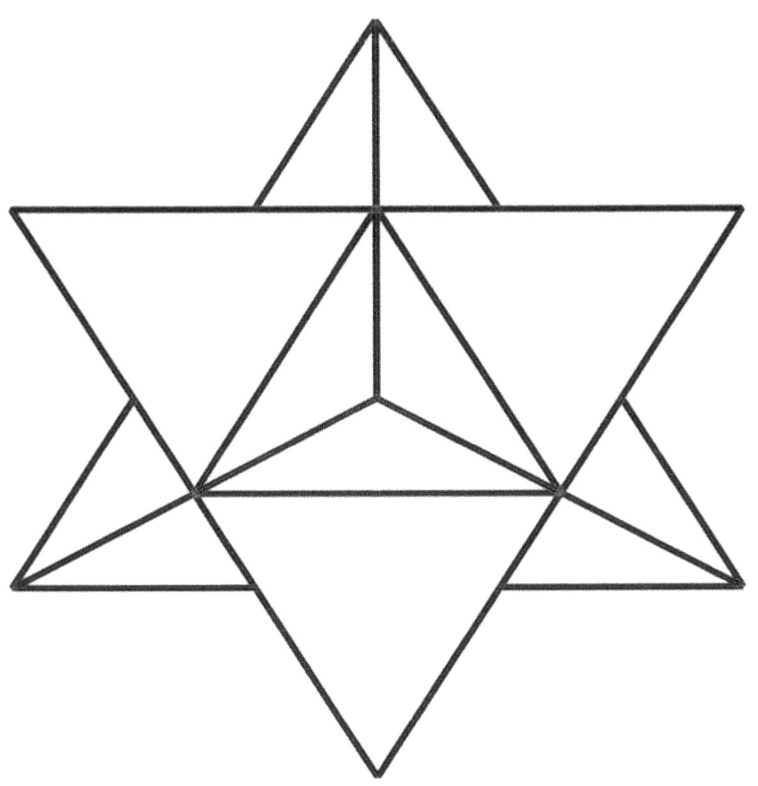

The Impact of Soul Intelligence® on Functional Medicine

The SOUL INTELLIGENCE® method is not just a tool for personal transformation; it is a profound complement to the field of functional and integrative medicine. At its core, SOUL INTELLIGENCE® allows us to address the root cause of many physical ailments—trapped emotions.

These emotions, when suppressed or left unresolved, manifest in the body as dysfunction, dissonance in frequency, or ultimately, disease. Our bodies are incredibly intuitive, always signaling us when something is out of alignment, acting as a barometer of our mental, emotional, and spiritual states.

In the realm of integrative medicine, the focus is on treating the whole person—not just the symptoms of a disease, but the underlying causes that give rise to it. This approach aligns perfectly with the principles of SOUL INTELLIGENCE®. The body is not just a physical entity; it is interconnected with the mind, emotions, and spirit. To truly heal, we must address all these aspects simultaneously. This holistic perspective is what makes SOUL INTELLIGENCE® such a valuable ally in the field of functional medicine.

When we consider how the body functions, we understand that every part of it, except for red blood cells, produces proteins. These proteins are expressions of our genetic code and can either upregulate genes for health or downregulate them for disease. If we don't address the internal conditions—our emotions, beliefs, and energetic patterns—we may only achieve temporary relief. The underlying issue will persist, leading to recurring symptoms or new health challenges. Functional and integrative medicine aims to uncover and treat these root causes, just as SOUL INTELLIGENCE® seeks to release the trapped emotions and energies that contribute to illness.

The idea that "your body keeps the score" is central to both functional medicine and SOUL INTELLIGENCE®. Trauma, whether from past experiences, generational patterns, or even spiritual imprints, is stored in the body. These imprints can manifest as physical symptoms, emotional blocks, or mental patterns that keep us stuck in cycles of pain and disease. By addressing these imprints through SOUL INTELLIGENCE®, we can help the body heal more fully and prevent the recurrence of illness.

In this context, we understand that healing is not just about treating the physical body. It involves a deep exploration of the mental, emotional, and spiritual aspects of our being. As we grow in awareness and understanding of our internal landscape, we can shift these patterns more quickly, leading to profound healing and transformation.

Case Studies of Healing

The power of SOUL INTELLIGENCE® is best illustrated through the real-life experiences of those who have used this method to heal. Here are a few examples:

1. **Healing Chronic Pain**: A client who had been suffering from chronic back pain for years, despite trying numerous treatments, discovered through a SOUL INTELLIGENCE® session that the root cause of her pain was tied to unresolved grief and guilt from a past relationship. By releasing these trapped emotions, she experienced a significant reduction in pain, and over time, it completely disappeared.

2. **Overcoming Anxiety**: Another individual struggled with debilitating anxiety that traditional therapy and medication couldn't fully alleviate. During a SOUL INTELLIGENCE® session, we uncovered deep-seated fears rooted in childhood trauma. By addressing these fears at the energetic level, the client's anxiety decreased dramatically, and she was able to reduce her reliance on medication.

3. **Addressing Autoimmune Disorders**: A healthcare professional suffering from an autoimmune disorder found that while functional medicine had helped manage her symptoms, the real breakthrough came when she addressed the emotional trauma connected to her condition. Through SOUL INTELLIGENCE®, she was able to release the emotional patterns that were contributing to her immune system's dysfunction, leading to improved health and well-being.

4. **Relationship Stress:** An executive divorced her husband and pursued a relationship with her business partner. While

on one level she knew this wasn't an appropriate relationship, she was drawn to him energetically because he was a soul mate. This deep bond pulled them together in spite of what society would tell us was morally wrong. Their relationship was layered and complicated as they were business partners and attempting to quietly be a couple while running a company. They were on again/off again, and this made her feel abandoned emotionally, not important, more like a mistress than a partner. She was also suffering with gastrointestinal issues and hormone imbalance, and her immune system was compromised.

We used the SOUL INTELLIGENCE® method to help alleviate some of her physical distress while working on the emotionally deeper issues that were being reflected in her relationship. She has since shifted her lack of self-confidence and gotten her groove back. She has orchestrated a successful exit professionally and is pursuing a new relationship personally where she is at the top of someone's list of importance.

5. The Battle of Good & Evil Within Oneself: A doctor who was also a professional athlete, athletic trainer, and wellness professional to high-net-worth individuals was plagued with physically debilitating issues consistently. He had a stable of healers who worked on him for years, where he experienced some relief but not with sustainable results. He was referred to me by another doctor. We began using the SOUL INTELLIGENCE® method to alleviate some of his physical discomfort which ranged from the inability to climb stairs and throbbing pain in his chest, to digestive issues and muscle spasms throughout his body. Working together, he was able to experience immediate relief of the physical issues.

The challenge was in finding the root cause of his issues. (This is where it gets a little weird.) Dark energy exists as does light energy. His mother was particularly sadistic and abused him even as a baby. His father was dismissive and not supportive. Beyond that, he had experienced lifetimes of being a part of the dark arts, and here he was in this lifetime practicing as a healer dedicated to the Divine. Through the power of SOUL INTELLIGENCE®, we were able to dig deep to discover how the energy from these other timelines continued to affect him in this life. Since our work together, his physical well-being has dramatically improved and he continues to work with me on a regular basis as part of his ongoing health regimen.

6. Spousal Guilt: Another healthcare professional was in the beginning of a very difficult divorce, truly struggling with her faith and her decision to leave the abusive marriage that she was in. She openly shared that her future ex-husband had suffered with pancreatic cancer and that she also bore the guilt of leaving him while he was recovering.

7. I shared with her that the root cause of pancreatic cancer is someone who has little time for fun and pleasure—duty and professional activity are more important. Their lives lack joy. They are regimented, serious people who feel a tremendous injury was done to them in the past. They feel it was profoundly unfair because rules were broken. The rules they are devoted to. They suppress the anger, doubt, and confusion and won't verbalize it. They carry on grinding away and being eaten by their own digestive enzymes. Needless to say, she was blown away as her husband suffered physical abuse as a child and never learned how to recover from that. She further realized that this was his burden to carry, not her issue to bear and problem to solve.

We used the SOUL INTELLIGENCE® method to help her feel more empowered to leave her relationship and have the confidence, strength, and courage to leave this abusive situation. She is now happy, in a new career, and in a new loving relationship.

8. One more try: Ever feel like you need to give a situation just one more try? Well, like Yoda says, don't try, do. Most of the time, this is our head trying to rationalize what our heart is trying to tell us. We know we don't really want to do it, but it seems like the right thing to do. We should try. We should do it. Well, this was exactly where one of my C-Suite executive clients was when we started working together. He was at the top of his game professionally, earning more money than he had ever in his life. The company was succeeding beyond all expectations. He loved his work, and his clients loved what his company produced and how he and his team supported them. But one thing was missing. He wanted his marriage and his relationship to be what he envisioned as loving, supportive, sensual, romantic, fun, and adventurous. All the things that he perceived as happy couples in love experience.

He didn't feel close to his spouse anymore. Sex was an act of obligation. Touch was infrequent. He felt like an ATM machine. We started working together on what was within him emotionally and energetically that was being reflected in his spouse. Using the SOUL INTELLIGENCE® method, we were able to get to the underlying root cause of his issues—unworthiness, self-limiting belief of not being good enough and feeling like you can't win, not important, invisible, and abandonment. Many of these old wounds stemmed from childhood and then became the lens through which he viewed the world, energetically attracting his spouse, who reinforced those old hurts.

After doing the work together, his spouse told him that it was the best year of their marriage. When he shifted, she shifted. The energy was different between them. In spite of this short-term success, they ultimately decided that they were better apart than together. Hence, trusting your SOUL INTELLIGENCE® rather than your head and even your heart, is the best option for avoiding additional pain and suffering. Trust your intuition, your inner knowing, your SOUL INTELLIGENCE®.

These case studies demonstrate the profound impact that SOUL INTELLIGENCE® can have when integrated with functional and integrative medicine. By addressing the emotional and energetic root causes of illness, we can facilitate deeper and more lasting healing.

Collaboration with Healthcare Practitioners

The integration of SOUL INTELLIGENCE® into healthcare practices offers significant benefits for both practitioners and patients. Healthcare professionals who embrace this method can provide more comprehensive care, addressing not only the physical aspects of health but also the mental, emotional, and spiritual dimensions.

For practitioners, SOUL INTELLIGENCE® offers a powerful tool to enhance their practice. It can be used alongside traditional treatments, functional medicine protocols, and other integrative therapies to help patients achieve more complete and lasting healing. By incorporating this method, practitioners can better understand the underlying emotional and energetic patterns that contribute to their patients' conditions, allowing for more targeted and effective interventions.

Patients, in turn, benefit from a more holistic approach to their care. They are empowered to take an active role in their healing journey, understanding how their emotions and energy influence their health. This not only leads to better outcomes but also fosters a deeper sense of well-being and personal empowerment. The collaboration between SOUL INTELLIGENCE® and healthcare practitioners represents a new frontier in medicine—one where the mind, body, and spirit are treated as an interconnected whole. This holistic approach not only addresses the symptoms of disease but also promotes true healing at all levels of being.

Journal Prompts

What is one relationship that you believe reflects a childhood wound for you?

Which of the case studies resonated with you? Why? What did itbring up for you?

CHAPTER 8

THE FUTURE OF SOUL INTELLIGENCE®

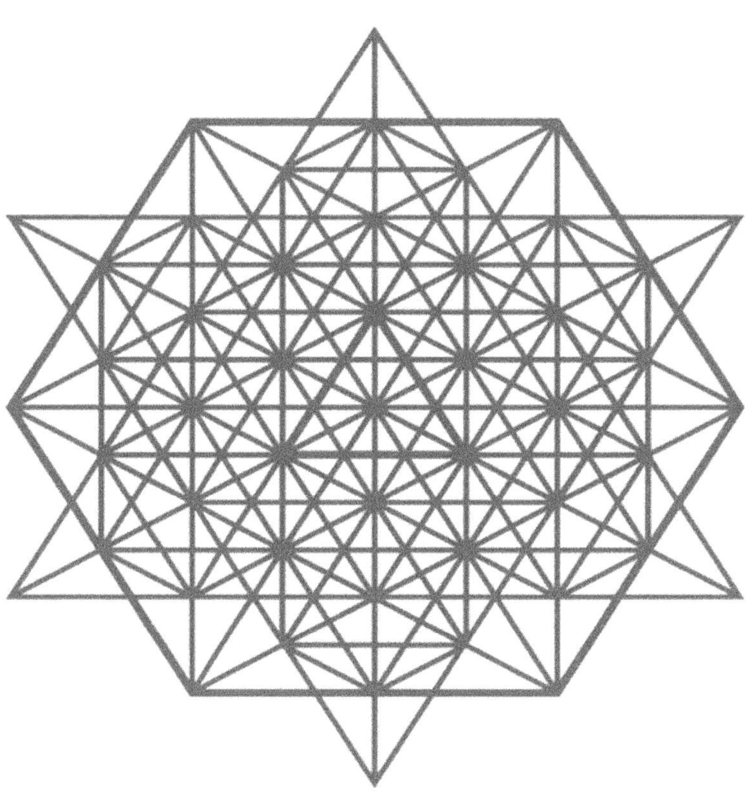

Group Healing and Collective Energy

As we explore the potential of SOUL INTELLIGENCE® in the broader context, it's essential to understand the power of group healing and collective energy. We are all energy beings, and our energy fields are constantly interacting, not just with individuals but with groups, teams, families, and entire organizations. Just as individuals have energy fields, so do couples, families, departments, divisions, and even companies. These collective energy fields can be influenced, cleared, and harmonized using the SOUL INTELLIGENCE® method.

This concept is rooted in what's known as the morphogenic field—a field of energy that is intensified when two or more are gathered. When clearing a group, we focus on the commonalities and shared energies that bind the group together. What issues are showing up collectively? How can we shift the energy to benefit everyone involved? I offer free monthly group clearings, often focusing on the prevailing energetic or astrological influences affecting us all. These sessions tap into the collective energy and provide a unique opportunity to experience the profound impact of group healing.

To grasp the essence of collective energy, think about walking into a crowded room —a cocktail party, a boardroom, or a rock concert. You can feel the energy, whether it's uplifting or draining. You sense the vibe without needing to think about it. What you're experiencing is the morphogenic field in action; it is the palpable energy that exists within a collective. You can absorb it, feel it, or send your own energy out and own the room—it's your choice.

Understanding Timelines —Past, Present, Future

Whether or not you believe in past lives, we all carry imprints from various timelines we've experienced. These imprints can manifest as negative patterns, qualities, or trends passed down through generations, much like genetic predispositions. We inherit not only physical traits but also habits, goals, archetypes, patterns, and the lessons our ancestors have attempted to work through. Sometimes, the challenges we face today are deeply rooted in the behaviors and beliefs of our family lineage. Perhaps a judgment or condition you are struggling to shift has been passed down through generations.

For example, one of my clients experienced ongoing issues with finances. She could generate income but not keep it. Another unexpected bill would arrive to take what she had just earned. During a SOUL INTELLIGENCE® session, I discovered that this was actually a generational issue that we could clear. When I clear something generational, we look to see how many generations back was the origination point and then clear it seven generations back and forward. She had the realization during our session that this was also a trait that her father had experienced, and his father before him. Recognizing this generational condition allows us to understand that we are powerful spiritual beings, having chosen to express ourselves in human form.

We are aspects of the Divine, navigating a human experience, and have the power to free ourselves from energetic and karmic debt for not just ourselves but for our families as well. As we become more aware of what's happening internally, we start to see how it reflects externally. Our growing awareness and understanding enable us to shift these patterns more quickly, leading to profound personal and collective transformation.

Holistic Integration of Mind, Body, and Soul

In functional medicine, the first place most doctors begin to treat is the gut because that is where most issues begin. According to Does Your Body Lie by Luis Martins Simoes, the stomach's first function is to serve as a recipient, to store food. The second function is to produce acids that attack and decompose the food. So, stomach problems represent a difficulty in allowing oneself to experience feelings and emotions, and secondly, a great difficulty in digesting them.

When the emotional conflict is intense, the stomach needs more gastric fluid to digest better. If a person attempts to swallow their anger, they swallow bile and feel the stomach acids in the form of heartburn. The biological reason is that to digest all of this, the stomach has to produce more gastric acid. Thus, gastric secretion is directly associated with the mind.

- Approximately 80% of the total immune system is located in the gut—within the mucosa of the GI tract, to be exact. The intestinal tract acts as a barrier to prevent harmful substances from entering the bloodstream. If this barrier is compromised, inflammation can occur, which can lead to chronic conditions.

- The gut produces 90% of the body's serotonin. Serotonin is the happy hormone, our natural mood booster.

- The gut microbiome is the collection of microbes in the gastrointestinal tract that helps maintain health. When the gut microbiome is out of balance, it can lead to chronic inflammatory bowel diseases, metabolic disorders, and other conditions.

- The gut is also home to 100 trillion microorganisms, including bacteria, viruses, fungi, and protozoa.

Even Hippocrates said over 2,000 years ago, "All diseases begin in the gut." Disease with oneself, with others. Isn't that where most of our challenges in life begin? Energetically, the gut is associated with the third chakra, or energy center, located in the solar plexus of the body.

The issues related to self—how you feel about yourself, your self-esteem, how you are perceived by others, your level of self-confidence, and how you put yourself out into the world—are all associated with this center point in the body. The old expression, The way to a man's heart is through his stomach, is definitely not just about cooking anymore!

Leaky Gut, Leaky Brain and More

There are numerous studies about the importance of gut health—one that jumps out is "Marital Distress, Depression, and a Leaky Gut: Translocation of Bacterial Endotoxin as a Pathway to Inflammation" from Psychoneuroendocrinology. In short, researchers drew blood from couples during marriage counseling, and the result was that participants with more hostile marital interactions had higher LBP (lipopolysaccharide) than those who were less hostile, meaning they were experiencing inflammation in the gut and an increase in gut permeability, or leaky gut syndrome.

Another one from PubMed states: All disease begins in the (leaky) gut, resulting in chronic inflammatory diseases. One article in the American Journal of Psychiatry states that there is evidence of gastrointestinal problems in autism and ADHD patients. A significant number of ADHD patients also experienced IBS (irritable bowel syndrome).

And here are a few more:

- Leaky gut syndrome: 30x risk of autoimmunity. International Journal of Molecular Sciences

- The brain-gut connection... Hopkinsmedicine.org

- Is your brain leaking? Translational Psychology 10

- Leaky gut causing brain health issues... ncbi. pmc5440529

- Signs you might have a "leaky brain"...National Institute of Mental Health

- Is your brain leaking? Trinity College Dublin Postmortem

- Leaky brain in neurological and psychiatric disorders. Sage Journals Australian

- Long COVID/leaky gut/leaky brain. Alzheimer's Journal

Leaky gut causes inflammation damaging cells in the digestive tract, causing gaps where undigested foods, pathogens, and toxins leak through the gut lining into the bloodstream. Due to unhealthy dietary and lifestyle choices, environmental toxicity, stress, and gut flora imbalance, your gut can become permeable or leaky. This means food particles and toxins can pass through your gut lining, and get into your bloodstream, stimulate your immune system, increase inflammation, and increase your risk of disease.

Leaky gut affects the whole body as it leads to inflammation throughout our systems.

- Digestive symptoms–bloating, diarrhea, pain, stomach cramps after eating, food sensitivities.

- Brain–migraines, headaches, depression, anxiety.

- Sinus and mouth–frequent colds, sinusitis, and food sensitivities.

- Joints-Pain, Rheumatoid Arthritis, Fibromyalgia. Adrenals – fatigue, chronic.

- Weight Gain

- Thyroid–Hashimoto's, Hypothyroidism, Graves. Colon–constipation, diarrhea, IBS.

There is another condition that is associated with leaky gut, called leaky brain. It includes blood-brain barrier dysfunction and symptoms include brain fog, neurofatigue, motor delay, clumsiness, anxiety, depression, poor memory, and personality changes.

When you consider how everything begins with our ability to process our feelings and emotions, it's no wonder our gut is typically the key to optimizing our physical health. More specifically, what happens when you resist dealing with your feelings and emotions? It further illustrates the power of emotions and what happens when energy pools, causing inflammation or swelling of the tissue of the stomach lining. The correlation between physical well-being, our emotional health, and our mental state is undeniable.

Vision for the Future of Soul Intelligence® in healthcare

The future of SOUL INTELLIGENCE® extends beyond personal healing into the realm of global healthcare. I envision a network of holistic healthcare practitioners across the country—and eventually worldwide—using this method to transform the way we approach health and wellness.

The future of medicine is frequency. All emotions we experience have a frequency and can be measured in hertz. There is no organ system in the body that is not affected by sound, music, and vibration. Disease can also be experienced as a form of disharmony, dysfunction, or a dissonance in the frequency. SOUL INTELLIGENCE® works on addressing those trapped energies, emotions, and dissonant frequencies and shifting them up and out of the physical, mental, emotional, and spiritual bodies.

The future of SOUL INTELLIGENCE® extends beyond personal healing into the realm of global healthcare. I envision a network of holistic healthcare practitioners across the country—and eventually worldwide—using this method to transform the way we approach health and wellness. In fact, I envision holistic quantum healing centers that combine quantum physics principles with holistic healing practices, emphasizing the connectedness of all aspects of the self. SOUL INTELLIGENCE® would be positioned as a key tool in this paradigm offering a cutting-edge approach to healing.

My goal is to develop a train-the-trainer program where certified practitioners can teach others the SOUL INTELLIGENCE® method in workshops and seminars. I see mental health counselors and psychologists using this approach to help patients get to the root cause of their traumas and trapped emotions, providing a deeper level of healing than traditional methods alone.

I plan to travel the world, speaking at holistic healing summits, seminars, and conferences, spreading the message of SOUL INTELLIGENCE®. I envision hosting holistic healing weekends and week-long retreats in high-vibrational, exotic locations around the globe, creating spaces where individuals can immerse themselves in this transformative work.

This is about more than just a method—it's about creating a global movement, an international SOUL INTELLIGENCE® community dedicated to uplifting the consciousness of the planet for generations to come. An exploration of the potential for group healing and the collective energy field in the context of Soul Intelligence®.

Journal Prompts

What changes for good are you willing to make within yourself to help raise the consciousness of the planet?

Have you ever experienced gut health issues? If so, think back to the circumstances that may have created them? Can you remember an emotionally charged event that could have startedthe physical discomfort?

CHAPTER 9

TRAINING AND CERTIFICATION

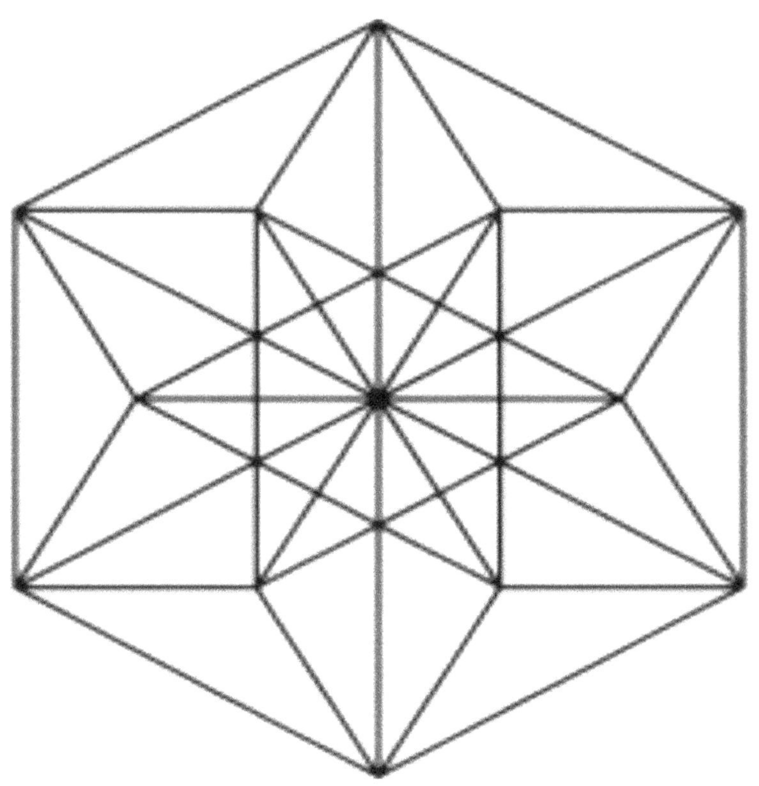

Becoming a Soul Intelligence® Practitioner

If you have a passion for healing and helping others—whether you are a healthcare professional, healer, coach, or simply someone who wants to make a difference in your own life, or for your family or community—you are the ideal candidate to become a SOUL INTELLIGENCE® practitioner. This practice is for anyone who feels called to bring more healing and alignment into the world.

As with most things that align with our soul's purpose, the universe conspires to provide everything you need when you are ready to pursue this path. The SOUL INTELLIGENCE® certification course is available online and is evergreen, which means you can begin at any time. You can visit www.soulintelligencemethod.com to sign up and get started on this transformative journey.

Once you have registered and paid for the course, you gain access to the educational portal, where all course materials, including downloadable charts, are at your fingertips. The only materials you need to gather on your own are a three-ring binder, plastic sheet protectors, and a pendulum no longer than the length of a pen. I encourage practitioners to personalize their binders as they prepare for their journey, organizing their charts to see how they flow from one to another.

One of the first skills you'll develop is working with your pendulum, learning how to trust the information it provides. I teach you how to program your pendulum by asking your Higher Self to guide its movements—a "yes" response is a backward and forward motion, a "no" is stillness, and a working circle clears energy. This connection between you, your pendulum, and the Divine is the cornerstone of the SOUL INTELLIGENCE® method.

It's important to keep your pendulum short (about the length of a pen) for ease of use, as you will be holding it over the charts for an extended period. The charts themselves resemble a pie chart, with each section representing specific information about the client. As you work, the pendulum will indicate what needs to clear, typically by moving clockwise. If the movement changes, it may signal an interference that requires attention.

Trusting the process is key to becoming a successful practitioner. You must learn to get out of your own way and allow the guidance from the Divine to come through, even if it doesn't make sense at first. Each session can feel like solving a mystery, and the accuracy of the information often amazes clients. The real joy comes from the feedback you receive during sessions—clients are often blown away by how much you know about them without ever having met them. This validation builds confidence in your ability to facilitate profound healing.

The Certification Process

The SOUL INTELLIGENCE® method is a revolutionary healing system that addresses the root cause of chronic conditions by releasing trapped emotions, trauma, and energy from the subconscious mind. As a certified SOUL INTELLIGENCE® practitioner, you can improve patient outcomes and enhance your practice by offering a powerful new healing modality that requires zero compliance—patients just need to show up and receive the healing.

The SOUL INTELLIGENCE® online course consists of the following modules:

1. Introduction to the SOUL INTELLIGENCE® Method

2. Overview of the Charts

3. Techniques to Uncover Blockages

4. Advanced Techniques to Release Deeper Issues

5. Exploring Session Types

6. Professional Practicum Certification

Throughout the course, you will learn how to work with a bioenergetic methodology that helps patients achieve lasting results. Once certified, you can offer this transformative healing system to your clients, adding value to your practice and creating a new revenue stream.

To complete the certification process, students must submit six different types of SOUL INTELLIGENCE® sessions for evaluation. These sessions can be recorded via Zoom or performed live with a certified SOUL INTELLIGENCE® practitioner coach. Sessions vary in focus and may include chakra clearings, pet clearings, physical ailments, parental clearings, generational clearings, and even space clearings. Advanced session types can address entity removals, crossing souls over, or clearing collective energy.

Once the sessions are reviewed, feedback is provided, and upon successful completion, you will receive a certification of completion, signifying that you have mastered the SOUL INTELLIGENCE® method.

Continuing Education and Community Support

The journey doesn't end with certification—continuing education and community support are vital aspects of the SOUL INTELLIGENCE® experience. By 2025, we will be launching an advanced SOUL INTELLIGENCE® practitioner course, designed to deepen your skills and knowledge. One new technique planned for this course is called Quantum Soul Mapping, where we map out your soul's journey across different lifetimes, identifying recurring patterns, archetypes, unresolved traumas, generational wounds, and your soul's ultimate mission. This technique would integrate elements of past life work, energy healing, and soul coaching. By identifying key experiences and lessons from previous lives, you can better understand your current life circumstances as they relate to aligning you with your soul's true purpose.

In the meantime, we have an active alumni group in our Slack community where graduates and students at all levels can connect, ask questions, and share their success stories.

We also offer weekly group coaching calls, providing 1:1 focused support and answering any questions about implementing the method with clients. This ongoing support ensures that you always have the resources and guidance you need to thrive as a practitioner.

Teaching and Spreading Soul Intelligence®

As we move forward, the SOUL INTELLIGENCE® method will continue to grow and evolve, with plans to expand its reach through a train-the-trainer program by 2026. This program will empower certified practitioners to teach the SOUL INTELLIGENCE® method to others, spreading this powerful healing system far and wide.

Imagine the ripple effect—each practitioner trained can go on to teach and heal even more people, creating a global network of healers who are aligned with their highest selves and making a profound impact on the world. The future of SOUL INTELLIGENCE® is one of expansion, empowerment, and collective healing, and you can be a part of this exciting movement.

Journal Prompts

Can you imagine making a difference for yourself, your family, friends and community using the SOUL INTELLIGENCE® method? If so, what would that look like? How would you make your world a better place?

How do you serve as a healer in your life? To your family, friends, community? What makes you feel good about living a life in service to others?

CHAPTER 10

PERSONAL HEALING JOURNEYS & TESTIMONIALS

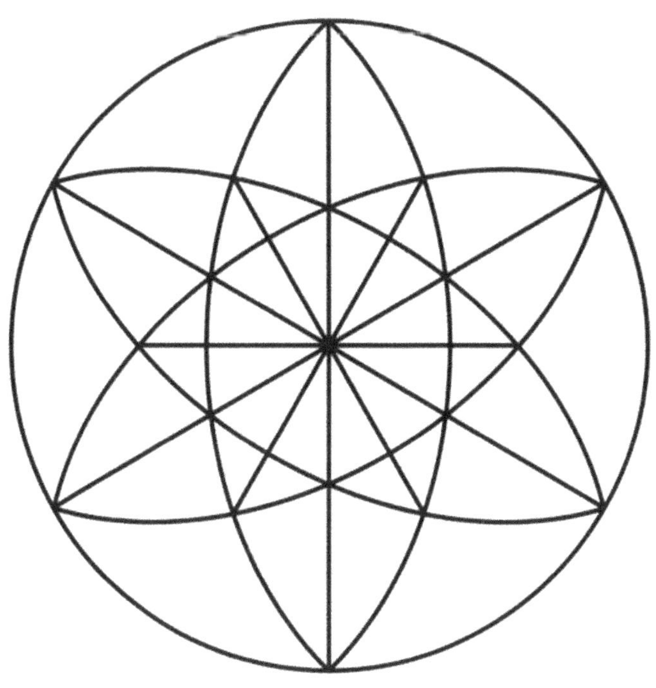

Stories from Practitioners

The SOUL INTELLIGENCE® method operates in many ways, like stem cells—going wherever the healing is needed. Whether it's mental, emotional, spiritual, physical, or energetic healing, this method addresses every level of your being. The client's experience shifts in multiple aspects of their lives, often in areas they didn't even realize were connected to their pain or struggles.

What I find remarkable is that the healing starts from the inside out. A client may no longer be triggered by the same old patterns that used to hold them back. Physical symptoms often clear up once the emotional component is acknowledged and shifted. Self-limiting beliefs, like the feeling of not being good enough, no longer drive people to overextend themselves. Old emotional wounds, like sexual trauma, lose their grip, allowing people to move forward with new relationships and new hope.

One of the greatest honors in my work is helping healthcare professionals. They are often seen as these all-knowing, unshakeable beings, but they, too, are spiritual beings having human experiences. Many healers enter their profession because they have healed themselves or walked through their own darkness. When they radiate their glow of transformation, they inspire others to heal. As the saying goes, "Healer, heal thyself."

A Doctor's Personal Transformation: Healthcare professional I worked with was struggling with her husband's interference in her practice. He was constantly pressuring her to move their practice to another state, and she was considering shutting down entirely due to frustration. Her physical body responded with chronic sinus infections and colds. Colds are mental confusion; sinus infections are related to being annoyed with someone you are close to.

After one session, she texted me the next morning, saying that she couldn't believe how different her husband was behaving towards her. She was awestruck that by doing a session for her, her husband began to react differently. She thanked me and shared that the session was absolutely incredible. She could hardly believe how much lighter, clearer, and more confident she felt. She then went on to expand her practice into two locations.

Clearing the Way for Healing: Another doctor had a foot issue preventing her from walking and a persistent cough she couldn't shake. I asked her what was going on in her personal life.

She shared that she was in the middle of a lawsuit and was feeling bullied by the opposing attorney. The foot pain, I explained, was tied to her inability to move forward, while the cough represented not being able to speak her truth. After one session, she contacted me, saying, "I want to thank you for our wonderful session yesterday, my foot pain is gone, and my cough is drying up. AMAZING!" She was also able to push the lawsuit forward, reclaiming her power in the situation.

A Profound Intervention: One of my clients, a nurse practitioner, performed CPR on a friend's husband, who unfortunately did not survive. The trauma from that event weighed heavily on her, leaving her depressed, angry, and feeling disconnected from her work. The details of the story are gripping and palpable.

While she was watching her friend's children play ice hockey, she was sitting with her childhood friend's husband. She went to get a drink at the snack bar and heard screaming coming from back near where they were sitting. Her best friend's husband had passed out and needed CPR. She performed CPR on him, giving him mouth-to-mouth while another bystander did compressions. He turned blue, was revived briefly, but unfortunately did not survive.

The Divine placed her in that situation for that "random" visit so that she could be there to support her friend through her husband's transition. She called me after it happened and scheduled a session for when she got home, as she did not feel like herself. She was depressed, angry, had lots of self-loathing and guilt, and didn't feel like she could work and see patients. When you exchange fluids or energy with another person, you can take on their energy, negative feelings, toxic emotions, etc. This is what happened to her on a subconscious level in trying to alleviate his suffering; she actually took on his anguish during his transition from body to spirit. This is truly done out of love and is not intentional.

After a SOUL INTELLIGENCE® session to separate the man's energy from hers, she texted me the next day, saying, "I feel so clear today and have hope. What you offer with SOUL INTELLIGENCE® is even more powerful than I realized. I am really honored to be able to speak about my experience and this incredible healing method you offer." She was able to heal emotionally and return to her patients with renewed clarity.

Stories of Transformation

Saving a Life, Literally and Figuratively

Another client, referred by her doctor, joined me on Zoom. As I worked, I sensed blockages in her heart and feelings of self-sabotage, self-destruction, and suicide. Gently, I asked if she had considered leaving this earth, and through tears, she admitted she had been neglecting her heart medication and nutritional supplements and contemplating suicide. She was very depressed and had been thinking about leaving this earth. She admitted it to me before I had to tell her what was coming up in the session.

She told me she didn't want to leave, so we continued the session, clearing the energy and reinforcing her commitment to life. I received her verbal agreement that she would resume taking her supplements and medications. I shared with her that I intended to share what came up, during the session with her doctor. She agreed that was a good idea. I told the doctor what came up and she called the client to review her symptoms to be sure that she didn't need to seek emergency medical care. The doctor thanked me for tag-teaming this client with her.

The very next day, both she and her doctor thanked me for the intervention. The doctor said she spoke with the patient and that she was feeling SO much better since our session. And gave me a BIG THANK YOU. The patient also told the doctor that she would take the nutraceuticals for her heart and stop self-sabotaging. The patient sent me several texts thanking me for the session and was filled with gratitude.

Midlife Crisis Breakthrough

Another doctor-referred client, who was going through her menopause, was feeling lost and unfulfilled. She didn't feel like herself any longer. She was looking at the life she's always known and realizing that she had been putting everyone else's needs, expectations of her, roles she's been expected to play—the dutiful wife, devoted mother, dedicated daughter—all ahead of her own spirit, her inner knowing, or as I know it—her SOUL INTELLIGENCE®.

She was miserable. She had sought solace outside her marriage, which on one hand made her feel more alive than she had in years, but on the other hand it only compounded her guilt and confusion. She was trying to honor her own evolution while still staying in an unfulfilling marriage. After just one session, she told me, "This has helped me more in one session than all my years of therapy." She was finally able to see her path forward, embracing her true self with clarity and compassion.

Chronic Illness and Emotional Healing

Another client comes to mind when I think about how powerful emotions truly are and what happens when you stuff them down inside, carry resentments, and ignore those feelings. Perhaps you were too busy to deal with them, your partner doesn't want to hear about it, or your family, friends, clients, career—all come first—you're last on the priority list.

Well, here's what happens when we do that. When emotions get trapped, they manifest as dis-ease within your body and can cause chronic illness. Louise Hay wrote a book, Heal Yourself A to Z, all about the root cause of disease based on emotional roots.

She wasn't honoring herself; she was always last on the priority list—kids, work, ex-husband, dogs. Then suddenly she started developing chronic conditions. By the time she was referred to me, she was on medical leave from her RN job for nearly a year with shingles, depressed, alone, under several doctors' care, and miserable. She was referred to me by her primary physician, as she had been suffering with a chronic illness for nearly a year and wasn't showing any marked improvement. She was also suffering with insomnia, torn up about her lack of a personal life, and really didn't want to go back to her stressful job. Before meeting me, she tried everything. I mean everything—medication, therapy, tapping, EFT, reiki, kinesiology, theta healing, acupuncture, ketamine, mindfulness, mushrooms, and more—without lasting results or real transformation.

She was desperate to move forward. Using the SOUL INTELLIGENCE® method, we figured out the root cause of her chronic illnesses. She immediately had relief from some of her symptoms. Her anxiety lessened, she began sleeping through the night again, and she woke up with more energy than she had in nearly a year. She started to heal her lack of self-forgiveness and met herself with true compassion.

What was the root cause? For her, it was deep-seeded fear, tension, dread of what's going to happen next, and being extremely sensitive to others' energy. She was very empathic and in a demanding job with an unappreciative boss, traveling weekly, and in hospitals around sick people all the time.

The way we shifted it was by releasing the energy that was stuck in her mental, emotional, physical, and spiritual bodies. The traditional coaching route would say, Get a new job! Well, she wasn't in a position to leave her job, so coaching her around making a career change wasn't an option. Often, we think what needs to change is outside of us, when what really needs to shift is what's within us. We need to uplevel from within.

She and I went even deeper within as our next step. We were able to release old blocks, self-limiting beliefs, and past traumas up from her subconscious mind to her conscious mind—as far back as childhood—without her having to "relive" the pain of the underlying events that caused it.

After working together, she made a full recovery and is back to work. I also provided her with tools to protect herself energetically in those challenging environments. She is now energized, happy, in a different role at her company, and pursuing a new personal relationship.

A Breakthrough in Business

One client, a successful coach, was attracting negative clients who didn't appreciate her work. Despite being positive and authentic, she found herself dealing with ungrateful clients who demanded refunds and left hurtful comments.

She was hurt—devastated really. She felt like she was expending all her energy, being positive, and sharing her program in a very authentic way. So, to have unappreciative, even hurtful clients join her program? She just couldn't understand what was happening.

After some hesitation, she decided to invest in a SOUL INTELLIGENCE® session. She was concerned over making the investment of both time and money. Stressing about how long it might take and how much money it might cost. Then she reflected that reworking her program hadn't worked. Changing her ideal client hadn't worked. Using new testimonials hadn't worked. She "tried all the things" to no avail. When she realized that she was worth it, she decided to invest in herself, and we scheduled a session. We discovered the dissonance in her frequency. It was like a broken radio antenna—some of the information was getting in but it was distorted—off.

We were able to identify why she was attracting the wrong clients— they were reinforcing several old self-limiting beliefs that whatever she did wasn't good enough, that she was unlovable, and over-the-top, just too much, "extra. "Little did she know that her insecure 5-year-old self was still running the show, even though her adult self had created a successful business. After our session, she reported a complete shift in the type of clients she attracted, finally aligning her business with her true mission and values.

Healing Through Connection

Another client, a young woman, was referred to me by her doctor. She was dealing with deep depression and loneliness, questioning her life choices and feeling unworthy of love. Through our sessions, she began to see how her past traumas and limiting beliefs were preventing her from fully connecting with others.

We worked through her self-doubt, and she left our sessions with a renewed sense of purpose and an open heart, ready to embrace the relationships and opportunities coming her way.

Impact Stories: Healing and Transformation

A Discussion with Dad

A brother and sister approached me about doing a session after their father had passed away. They both wanted to know that he was OK and that he hadn't suffered. Up until this session, I had never considered myself a medium. However, when you consider that the information comes through me—I suppose that is the definition of a medium! I shared with them that sometimes people do come through, but the way I discern the information is using the charts to piece together the messages.

My primary sense is clairsentience, meaning I feel energy. When I feel a being is in my field, I feel pressure on the top of my head. Some people have the gift of sight, which is known as clairvoyance; some hear, which is called clairaudience; and still others experience claircognizance—an inner knowing. We all have access to all the "clairs," but typically, just like we have a propensity for certain skills over someone else, we have a primary sense.

I told the siblings that I would set the intention of the session to connect with their father. Sure enough, he came through. As I was divining the message and working to answer their questions, I recall saying to them, "Wow, your dad is really chatty for a guy." They both laughed really loud and said, "Yes, that's totally him! "After our session, they both experienced a deep sense of peace and shared the messages with their mother. It's really satisfying knowing you provided comfort to a family after a sudden death like they experienced.

Eating Disorder

I was approached by a mom who was really distraught over her young daughter's inability to lose weight. She put her on a healthy diet, vitamins and good nutritious snacks, and tried hypnotherapy, exercise, and counseling. Nothing she did seemed to have an impact on her daughter's health. Childhood obesity is growing in our country, and it's usually due to the lack of nutrients in our food and poor choices of sugary and hormone-filled junk. This young girl was doing all the right things and still couldn't get to a healthy weight. So, we did a SOUL INTELLIGENCE® session together. We found that in a past life, she had starved to death. So, in this life, she always ate more than she needed as her body remembered that trauma. Once we did the session, she was able to curb her appetite and lose weight.

Couples Counseling on a Higher Level

One of my favorite things to hear after a session is how someone's spouse changed and shifted when we did the energy work on only one of them. It gets really interesting doing the work for both halves of a couple. I recall doing a session for a husband who had such a profound experience that he suggested his wife also experience a session. Not only did we work on their individual challenges, but we also cleared what they were reflecting back to one another in terms of their marriage. The unresolved hurts, the unmet needs, the feelings that they were repressing for the sake of "keeping the peace."

We also cleared the energy field around them as a couple, so collectively what needed to shift. They told me that this was better than "Marriage Encounter," which was a Catholic program that they participated in to help strengthen their marriage. Where they once were contemplating separating prior to working together, they are now even more deeply connected as a couple and planning a trip abroad to celebrate their renewed love for one another.

Brotherly Love

One client was experiencing pain in his right arm. He couldn't understand why it was hurting, as he works out on a regular basis, is in great physical shape, and didn't recall pulling anything while exercising. I shared with him that the right side of our body represents male energy. Given that it was his right arm, it indicated that he was storing stress, pain, and limitation from a male in his life. We performed a SOUL INTELLIGENCE® that showed an issue that was coming up for his brother. He is very close to his brother, and his brother was struggling in the throes of a divorce. He realized that he had been taking on his brother's stress and anxiety. Once he acknowledged it, the pain subsided.

These stories highlight the transformative power of the SOUL INTELLIGENCE® method. Whether it's healing physical pain, clearing emotional trauma, or shifting limiting beliefs, this method offers profound change at every level. The feedback I receive from clients is a constant reminder of the power of this work—how aligning with our SOUL INTELLIGENCE® allows us to heal, grow, and live in our highest truth.

Journal Prompts

If you could write your own case study following a **SOUL INTELLIGENCE®** session, what would it say? What would you want to shift physically, mentally, emotionally, spiritually, energetically?

If you could give a SOUL INTELLIGENCE® session to someone you know, who would it be and why? What would you want to heal them? How would their healing impact you

CHAPTER 11

THE VISION FOR THE FUTURE

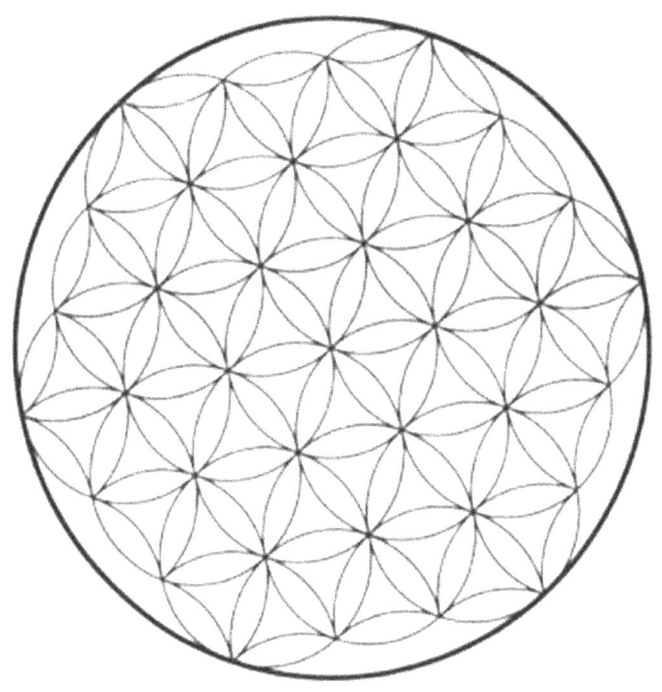

Creating a Global Movement

By 2025, my vision is to be leading the SOUL INTELLIGENCE® movement full-time, traveling across the country, sharing the power of this method, and teaching others how to integrate this incredible biohack into their practices, businesses, and lives.

The ultimate goal is to create a global SOUL INTELLIGENCE® movement that empowers everyone to heal completely and feel whole within themselves. I envision a world where individuals are balanced in their Divine masculine and feminine energies, fully aligned with their true power. This movement will raise the consciousness of the planet, one person at a time.

The vision for SOUL INTELLIGENCE® has always been about something bigger than individual healing. It's about creating a global movement—a shift in consciousness that extends beyond the personal to the collective. Each of us has a role to play in raising the vibration of the planet, and it begins with our own inner work.

As more people align with their SOUL INTELLIGENCE®, the ripple effects are felt across families, communities, workplaces, and eventually, the entire world. This movement is about empowering individuals to reclaim their innate power to heal, grow, and contribute to the greater good. It's about creating a new paradigm where holistic healing, emotional intelligence, and spiritual alignment are at the forefront of how we live and lead.

Imagine a world where everyone is operating from a place of deep inner knowing, fully aligned with their mission, vision, values, and purpose. A world where we make decisions from our SOUL

INTELLIGENCE® rather than fear or ego, where we work together to elevate the consciousness of humanity. This isn't just a dream—it's a reality we can co-create, and it begins with each of us choosing to step into our power.

The butterfly effect teaches us that small shifts in one person can create waves of change across the globe. In chaos theory, the butterfly effect is the sensitive dependence on initial conditions in which a small change in one state of a deterministic nonlinear system can result in large differences in a later state. In other words, when you make a change within yourself, it creates ripples that influence the collective consciousness. We are all interconnected, all originating from the same Source.

By working on yourself —by aligning with your SOUL INTELLIGENCE® —you contribute to the collective awakening. You become part of a movement that is not just about healing individuals but about raising the consciousness of the entire planet. When you embark on a journey of personal growth, you're not just changing your own life–you're creating a ripple effect that touches everyone around you. Your own transformation can uplift, inspire your family, friends, and community. By stepping into your highest potential, you become a beacon of hope and positivity, influencing others to do the same. The collective impact of this is powerful: as more individuals awaken to their true selves, communities become stronger, more compassionate, and more resilient. As we work on ourselves, our families, and our communities, we contribute to the global shift towards more peace, love, and harmony.

We are now officially in the Aquarian Age, which in its highest form means we are all to be in alignment with our true authentic self, on our Divine mission of service, while collaborating with others, and looking for ways to contribute to the collective whole. By focusing on healing, empowerment, and spiritual awakening, you're not just improving your own life —you're playing a crucial role in the upliftment of humanity. The potential for positive impact is immense: as more people embrace these practices, we will see a shift towards a more compassionate, conscious, and harmonious world.

Between 2023 and 2024, Pluto moved from Capricorn into Aquarius and established itself in Aquarius until 2043, which is said to have activated a powerful, essential societal transformation. Along with this transit, humanity takes a significant step into the Age of Aquarius and begins to settle into it. Some say that 2024 will be a year of new beginnings, revolution, and becoming more autonomous. Others say that it will be a year when Aquarians recognize their ability to bring about positive change.

The Age of Aquarius is anticipated to be a time of intellectual renaissance—with a lightning-fast exchange of ideas and whispers of wisdom floating in the air. It's also expected to be a time when people become more aware that their behavior affects other living beings and things. It is also anticipated to be a paradoxical time that highlights both the authentic, individual contribution and, at the same time, how that individual contributes to the whole in a very transparent way. I have this inner knowing that together we have incarnated in the perfect time where we can create a world that is more peaceful, compassionate, and connected.

My hope is that SOUL INTELLIGENCE® becomes as widely recognized as emotional intelligence was just a decade ago. I want it to spread like wildfire, empowering people to heal themselves and others holistically.

While some have described it as "Reiki on steroids," SOUL INTELLIGENCE® is much more specific and powerful, directing energy in ways that are uniquely tailored to each individual's needs.

Imagine what we could achieve if everyone on the planet were clear, focused, and aligned—physically, mentally, emotionally, and spiritually. The potential for global transformation is limitless.

The Vision for Soul Intelligence® in Healthcare and Beyond

The future of SOUL INTELLIGENCE® is limitless, with applications that extend far beyond personal development. In the healthcare industry, the potential for transformation is enormous. Imagine a world where healthcare practitioners not only treat the physical body but also address the emotional and energetic root causes of illness. This is the future I envision—one where SOUL INTELLIGENCE® is integrated into healthcare systems, helping practitioners get to the root of their patients' suffering and providing holistic solutions that address the whole person.

The method has already begun to make an impact in functional and integrative medicine. Doctors, nurses, and therapists are incorporating SOUL INTELLIGENCE® into their practices, helping patients achieve deeper levels of healing than ever before. By releasing trapped emotions, clearing energy blockages, and aligning with their highest selves, patients are experiencing profound shifts in their health and well-being.

Holistic Quantum Healing is the future of wellness, recognizing that every thought, emotion, and physical sensation is interconnected at the quantum level. Through SOUL INTELLIGENCE®, you can tap into this vast network of energy and information and transform your life on all levels. As I've shown, SOUL INTELLIGENCE® is a pioneering, innovative approach in the wellness and spiritual healing space. I envision Holistic Quantum Healing Centers where people receive treatment and healing on all levels—physically, mentally, emotionally, spiritually, vibrationally, and energetically.

But the potential for SOUL INTELLIGENCE® doesn't stop with healthcare. The business world, too, is beginning to recognize the value of operating from a place of emotional and spiritual alignment. Companies that embrace these principles are seeing improvements in productivity, employee satisfaction, and overall success. As more leaders tap into their SOUL INTELLIGENCE®, they can create workplaces that are not only successful but also supportive, compassionate, and aligned with higher purpose.

In personal development, the SOUL INTELLIGENCE® method offers a pathway for individuals to break free from limiting beliefs, heal old wounds, and step into their full potential. Whether it's overcoming past trauma, building stronger relationships, or achieving professional success, this method provides the tools and guidance to create lasting transformation.

Conclusion: Join the Movement

The SOUL INTELLIGENCE® movement is more than just a healing method—it's a call to action. It's an invitation to step into your true power, to heal yourself, and, in doing so, to contribute to the healing of the world.

We are at a pivotal moment in human history. As more people awaken to their true potential, we have the opportunity to create a new way of living—one that is based on love, compassion, and unity. Each of us has a role to play in this global movement. By aligning with our SOUL INTELLIGENCE® and living from that place of deep inner knowing, we can create the world we've always dreamed of.

Now is the time to take action. Whether you choose to become a certified SOUL INTELLIGENCE® practitioner, incorporate the method into your personal or professional life, or simply share this knowledge with others, you are contributing to the collective awakening.

Join us in this movement. Together we can raise the consciousness of the planet, heal ourselves and others, and create a world where everyone thrives. This is not just the future of healing—it's the future of humanity. Let's build it together.

Journal Prompts

What is your vision for the future?

How would you like to contribute to the collective consciousness?

CONTACT INFORMATION

How to get in touch with the author, book sessions, or learn more about Soul Intelligence®.

Visit www.soulintelligencemethod.com to join my email list, access free resources including meditations, and my sacred soul script, access to free group sessions, and online webinars.

On my website you can also book a session with Kristine Genovese, some of my best certified SQ practitioners, and learn more about the SOUL INTELLIGENCE® CertificationProgram.

RESOURCES

Glossary

Bioenergetics: The study of energy relationships and energy transformations and transductions in living organisms.

Collective Consciousness: a sociological concept that refers to the shared beliefs, values, and norms that bind a society together. It's a unifying force that shapes how people behave, interact, and perceive each other in a community.

Esoteric Message: An esoteric message is a message that has a secret or hidden meaning that is only understood by a select few people who have special knowledge or interest. The word "esoteric" comes from the Greek word esōterikos which means "inner" or "further inside."

Morphogenic Field: The energy field around objects in space-time as "nonmaterial regions of influence extending in space and continuing in time." These fields contain a memory for the object or system they organize: a plant, an animal, even an atom or a snowflake.

Neurobiology: The biology of the nervous system.

Quantum Physics: the study of matter and energy at the most fundamental level. It aims to uncover the properties and behaviors of the very building blocks of nature. While many quantum experiments examine very small objects, such as electrons and photons, quantum phenomena are all around us, acting on every scale.

Quantum Soul Mapping: A technique that traces the soul's journey across lifetimes to identify recurring patterns and unresolved traumas to better align you with your soul mission.

Soul Resonance: The state of being when a person's actions, thoughts, and feelings are perfectly aligned with their soul's purpose.

Further Reading

If you are interested in reading other books related to Soul Intelligence®, energy healing, and consciousness, I recommend the following:

"Awakening into Oneness" by Arjuna Adagh.

"Loving What Is" by Byron Katie.

"The Power of Now" by Eckert Tolle.

"Way of the Peaceful Warrior" by Dan Millman.

"Whatever arises, love that" by Matt Kahn.

"Zero Limits Living" by Dr. Joe Vitale

Tools And Resources

A list of tools, websites, and resources to support readers in their journey with Soul Intelligence®.

"Does Your Body Lie? Heal the Person, not the Sickness" by Luis Martins Simoes

"The Secret Language of Your Body, The Essential Guide to Health and Wellness" by Inna Segal

"Rise Sister Rise" by Rebecca Campbell.

Rebecca Campbell also offers guided meditations that I enjoy at www.rebeccacampbell.me

She also has a variety of card decks that I enjoy using for my own inner guidance and readings.

"You are a Goddess" by Sophie Bashford.

Sophie Bashford Meditation Journeys with Goddesses, Gods & Guardians available on Audible.

ABOUT THE AUTHOR

Kristine Genovese

I'm sure you're wondering, Who am I anyway? I'm Kristine Genovese, the creator of the most powerful bioenergetic method available today—the SOUL INTELLIGENCE® Method. But I wasn't always this brave, out-of-the-closet spiritual gangster.

For nearly 30 years, I served as a corporate growth and turnaround specialist, first in the field of higher education and more recently in nutraceuticals. My work focused on identifying people, processes, and technology that weren't functioning optimally, with the goal of streamlining operations and improving outcomes. Essentially, I would locate where the energy was stuck—whether in people, places, or systems—what was creating stagnation, and how we could shift that to create a better flow, whether that meant moving people into the right roles or implementing better technologies.

Today, as a conscious leadership executive, I help individuals and teams become the best versions of themselves—leading to greater performance, consistency, and alignment so they can make a bigger impact in their organizations and beyond. I use all of my abilities, including The SOUL INTELLIGENCE® Method, to positively impact the lives of my colleagues, clients, and their patients.

I bring to the table expertise in bioenergetics, marketing, sales, human capital development, operations, customer experience, and client retention. I hold a Master's in Organizational Management from the University of Phoenix and a Bachelor of Arts in Speech Communication from Bloomsburg University. I'm also an ICA Certified Professional Coach and a Certified Practitioner of the SOUL INTELLIGENCE® Method. I currently live in Tampa, Florida, and am the proud mother of two grown men—one a mental health nurse practitioner, the other an underwater bridge inspector. When I'm not working, I'm an avid tennis player, scramble golfer, recreational boater, weekend gardener, wine aficionado, and often found relaxing on the beach under a shady palm tree.